Just a Lark.

Just a Lark!

Text by Jim Flegg
Illustrations by
Norman Arlott
Photographs by Eric
and David Hosking

CROOM HELM
London • Sydney • Dover, New Hampshire

©1984 Text: Jim Flegg; photographs: Eric
and David Hosking; illustrations: Norman Arlott

Croom Helm Ltd, Provident House,
Burrell Row, Beckenham, Kent BR3 1AT

Croom Helm, Australia Pty Ltd,
First Floor, 139 King Street,
Sydney, NSW 2001, Australia

Croom Helm, 51 Washington Street,
Dover, New Hampshire
03820, USA

British Library Cataloguing in Publication Data

Flegg, Jim
    Just a lark.
    1. Birds — Dictionaries
    I. Title II. Hosking, Eric
    III. Hosking, David IV Arlott, Norman
    5989.03'31   QL673

    ISBN 0-7099-1049-5

Printed and bound in Great Britain

CONTENTS

Introduction 7
Adjutant 11
Albatross 12
Australian Magpie 13
Barbet 14
Bee-eater 15
Bishop 16
Blackbird 17
Blackcap 18
Bleeding-heart 19
Booby 20
Bowerbird 21
Broadbill 22
Bullfinch 23
Bunting 24
Butcher Bird 25
Canary 26
Cardinal 27
Catbird 28
Chanting Goshawk 29
Chatterer 30
Cisticola 31
Cockatoo 32
Cock-of-the-rock 33
Coot 34
Corden-bleu 35
Cormorant 36
Corncrake 37
Cowbird 38
Crane 39
Crossbill 40
Crow 41
Cut-throat 42
Dabchick 43
Dipper 44
Dove 45
Dowitcher 46
Eagle 47
Falcon 48
Firecrest 49
Flycatcher 50
Friarbird 51
Frigatebird 52
Frogmouth 53
Go-away Bird 54
Goose 55
Grackle 56
Grenadier 57
Guinea-fowl 58
Gull 59
Hammerhead 60
Hang-nest 61
Heron 62
Hobby 63
Honeyeater 64
Honeyguide 65
Hoopoe 66
Hornbill 67
Hummingbird 68
Ibis 69
Jay 70
Killdeer 71
Kingfisher 72
Knot 73
Lammergeyer 74
Lapwing 75
Laughing Thrush 76
Lily-trotter 77
Loon 78
Lorikeet 79
Lovebird 80
Lyrebird 81
Macaw 82
Mandarin 83
Martin 84
Meadowlark 85
Merlin 86
Mockingbird 87
Mollymawk 88
Mousebird 89
Myna-bird 90
Nightingale 91
Nightjar 92
Noddy 93
Nun 94
Nutcracker 95
Nuthatch 96
Owl 97
Oystercatcher 98
Paddy Bird 99
Painted Snipe 100
Parakeet 101
Pelican 102
Penguin 103
Pintail 104
Plover 105
Puffin 106
Quaker 107
Razorbill 108
Robin 109
Roller 110
Rosefinch 111
Rubythroat 112
Ruff 113
Saddleback 114
Sandpiper 115
Secretary Bird 116
Seedeater 117
Shearwater 118
Sheathbill 119
Shoebill 120
Shoemaker 121
Shoveler 122
Skimmer 123
Skua 124
Skylark 125
Snake-bird 126
Spoonbill 127
Starling 128
Steamer-duck 129
Stinker 130
Stork 131
Sunbird 132
Swallow 133
Swan 134
Tanager 135
Tern 136
Thicknee 137
Tinker-bird 138
Tit 139
Toucan 140
Treecreeper 141
Trumpeter 142
Turnstone 143
Vulture 144
Warbler 145
Waxwing 146
Weaver 147

Whalebird 148
Wheatear 149
Whipbird 150
Whistler 151
Whistling-duck 152

White-eye 153
Whitethroat 154
Whydah 155
Widow-bird 156
Willie Wagtail 157

Woodpecker 158
Wyrneck 159
Yaffle 160

Note: All caption lines giving English and Latin names refer to the photographs.

INTRODUCTION

Writing only 40 years ago, the late James Fisher listed among those whom he knew to be interested in birdwatching a 'Prime Minister, a Secretary of State, a charwoman, two policemen, two kings and one ex-king, five Communists, four Labour, one Liberal and three Conservative Members of Parliament, the chairman of a County Council, several farm workers, a rich man who earns two or three times their weekly wage in every hour of the day; at least forty six schoolmasters, and an engine driver'.

To these can be added today (reflecting both political and cultural shifts of emphasis) two Princes Consort, at least two Presidents, numerous Peers and Members of Parliament, Admirals, Field Marshals, and a vast array of people falling under the general heading of 'sports and show-business personalities'. The number of 'ordinary' birdwatchers has become impossible to estimate; the only guide to its enormity is the membership of the various local and national clubs and societies. In the United Kingdom alone, this has risen from 10,000–20,000 at the time James Fisher was writing to an estimated total over the half million mark, and of course there are many, many birdwatchers who, although not belonging to such organisations, share a genuine interest in, and feeling for, birds.

What, then, are the reasons behind this vast upsurge in interest in birds? What is their fascination? Perhaps it is because birds are one of the commonest forms of wildlife in our surroundings, always adding to the pleasure of being alive. So many and varied are they that all of us have ample opportunity to watch them. They occur in environments of all types, from remote mountain tops to the hearts of our cities, from sun-baked summer-holiday beaches to the gloomy depths of conifer woodland in winter. Many frequent our gardens, giving the extra privilege of allowing us to take advantage of their confiding nature and watch them go about their daily business at the bird table or nestbox.

Because of this, an enthusiasm for birds can make dull places interesting, routine trips enjoyable. Birds can enliven the walk to the station or to school, and can add another dimension to a holiday, either at home or abroad. A detailed investigation of the neighbourhood of your home can produce as many surprises and interesting discoveries as can a trip to a strange area, perhaps with a very different climate and vegetation. Often enough, this enjoyment can also be completely free of charge.

Like a kaleidoscope, birds change with the seasons. Spring brings songsters such as the Swallow and Nightingale back from Africa, and in autumn these are replaced by the equally thrilling, but visually more spectacular, wild geese and swans from the Arctic. Watching, marvelling at the beauty of plumage or listening to the variety of songs, leads on to other things. In one direction, birds can be seen in an artistic sense as just part of a wider landscape: the Skylark as essential an element of the summer sky as ducks flighting across the estuary mudflats in winter. In another, just watching birds leads to a deeper need to know more: about their behaviour, their nests, or about the spectacular feats of endurance and navigation that they perform on their enormous migratory journeys.

Thus the major reason for our enjoyment of birds seems to centre on their presence all around us, and on the unrivalled opportunities that they offer, allowing us simply to relish their beauty, just to watch their behaviour, or to study aspects of their lives with a time commitment that remains within our control – *usually* remains within our control, for birdwatching (like other hobbies) can 'hook' the participant in a more than normally compelling way! Not the least of the ways in which current interest has arisen has been through the efforts of the early pioneers in popular presentation of ornithology to a receptive public through books, radio and television. Though popular ('starch-free' would be an appropriate modern term), these presentations retained their accuracy, and those of us interested today owe a great deal to great names such as James Fisher, Sir Peter Scott, and of course Eric Hosking.

Birds offer, naturally, powerful stimulus to wildlife photographer, artist and author alike. Normally, the task of each is to document the various facets of birds and their lives, for example to help in their identification, or to illustrate their behaviour to a wider birdwatching or scientific audience, or to depict the birds, and their habitats, for purely artistic or aesthetic reasons. Those of us involved in the making of this book have each enjoyed a lifelong and deep personal interest in birds. We are privileged to be involved as professionals, with the dedication that necessarily applies to those studying often elusively wild creatures. But this does not mean (as is so often assumed) that our interest is confined to the earnest contemplation of the sombre and serious facts of avian ecology, or to the precise definition and portrayal (with pen, brush or camera) of each action or feather. *Enjoy* is the appropriate word, for birds are able to fascinate in so many ways, some of them less serious, even lighthearted. It is these aspects and their intrinsic intrigue and humour, gathered from the worldwide range of our collective experience, that form the basis of this book, a fun guide to some of the factually and figuratively most appealing of the world's birds.

ADJUTANT

Adjutant Stork *Leptoptilos dubius*

The dictionary definition of an adjutant — an officer assisting senior officers with the details, particularly administrative ones — implies a bearing both militarily stiff and upright and slightly clerical. With its ability to stand motionless and upright, sometimes on both legs 'at attention', others only on one leg, or squatting — presumably 'at ease' — for long periods, the Adjutant (a giant Indian stork) fits the first part of this definition. Just as soldier-like is its precisely-paced walk, which would make the Adjutant an ideal parade-ground figure in slow-march tempo. In its sombre grey plumage, and with its bald pinkish head fringed with a hair-like fluff of bedraggled feathers, it fits, too, the image of the subordinate clerk. Unfortunately the Adjutant's gluttonous feeding habits, and its to human eyes disgusting behaviour when squabbling noisily with vultures and others of its kind over the entrails of a long-dead cow, rather destroy the military image.

ALBATROSS

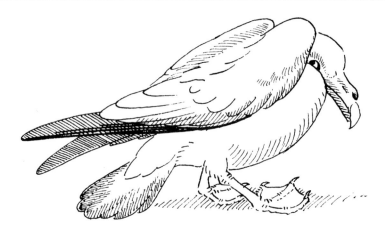

In the air, albatrosses on their long, slender wings are masters of their element, sweeping across the tropical oceans of the world on wings sometimes reaching 3 metres (10 feet) in span. Though superb in the air and exploiting every shift of wind and the slightest of upcurrents off the waves, albatrosses are incredibly — and often hilariously — clumsy on land. Their walk is a slow and ponderous — even pompous — waddle on huge feet, the webbing of one foot often being carefully placed on top of the other, making forward progress difficult. But it is landing that gives most trouble: slowing down in the air is difficult for them, especially on windy days, and many homecoming birds fail to arrest their progress before running into neighbours sitting on nests and finish up turning a series of ignominious somersaults. This may be one reason why the Shy Albatross got its name.

Shy Albatross *Diomedea cauta*

AUSTRALIAN MAGPIE

The Australian Magpie is one of a family of three slightly different species, separated by geography in that vast landmass and occasionally interbreeding where they meet. They are as pied in plumage as the Northern Hemisphere bird of the same name, but are crow-sized and lack the disproportionately long tail of their counterpart. All are fine singers, with a flute-like warble that puts them among the favourites of Australian birdwatchers. Like Magpies, though, they have their disadvantages and are disliked by farmers for their attacks on fruit and by some birdwatchers for their habit of preying on the eggs and young of smaller birds. Rather like mobs of teenage hoodlum humans, they form gangs which are strongly territorial, and during the breeding season become extremely aggressive in defence of their nests. The defence is usually by the males, which are aggressive enough to set fur flying and have marauding dogs or cats running off with their tails between their legs. They are often suburban birds, and may occasionally subject innocent humans to the same treatment!

Australian Magpie *Gymnorhina tibicen*

BARBET

The barbets are a colourful family of birds, mostly from tropical and subtropical areas of Africa and Asia. They have a catholic choice of diet, and a powerful jagged-edged beak that allows them to tackle most tasks, from excavating nest holes in living trees to eating fruit, seeds and flesh. Ornithologists initially anxious to find out more about their habits may tend to change their minds after handling a barbet and suffering bloodily savaged fingers. The Coppersmith Barbet's name originates in India, from the similarity of its 'tonk-tonk' calls to the tapping noises emanating from the small metalworking shops and stalls which are common features of the market place. Equally fascinating is another call, reminiscent of India, sounding just like a bicycle tyre being pumped up.

Coppersmith Barbet *Megalaima haemacephala*

BEE-EATER

The Bee-eater and others in its family are the rainbows of the bird world, some arrayed in plumages of such mixed colours as would seem extreme even on an artist's palette. They are skilful swooping fliers, hunting insects on the wing and as they pass on the ground. A common hunting technique is to choose an exposed vantage perch from which to sally forth after prey. As their name suggests, they make something of a speciality out of catching bees, wasps and the like, all equipped with powerfully venomous stings. Their horny beak is sting-resistant, but Bee-eaters are naturally careful by instinct and by parental tuition to return and batter their prey senseless on their perch before swallowing it. Only when the prey is as large, struggling and leggy as a praying mantis or stick insect do they seem to bite off more than they can chew.

Bee-eater *Merops apiaster*

BISHOP

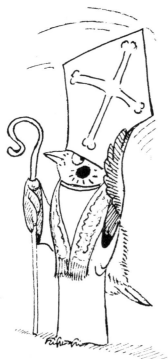

Bishops are birds of the grassy African plains. The crimson-and-black-plumaged Crimson or Red-crowned Bishop, hopefully unlike any similarly garbed ecclesiastical prelate, is polygamous, and may gather as many as six drab sparrow-like wives in the breeding season. Each of these he establishes in a domed nest in the long grass around his territory, and his life becomes one continuous round of patrols to deter intruders and ensure that his wives do not philander with neighbouring males. This vigorous defence of his territory and tribe of wives and children ceases as the dry season becomes established and the birds gather in flocks seeking seeds.

Crimson Bishop *Euplectes hordacea*

BLACKBIRD

Brewer's Blackbird is a member of the New World Icterid family, and, being Blackbird-sized and (roughly) Blackbird-shaped, it is not surprising that homesick emigrés from Europe likened it to their 'home' Blackbird and named it accordingly. Whereas the European Blackbird is a superbly melodious songster, little can be said about the song of Brewer's Blackbird save that it is scratchingly wheezy — akin to a gate swinging on rusty hinges. It is a fruit- and insect-eater, widespread in North America from Canada south to Mexico, a bird of open country and marshland that has taken well to suburban life and even penetrates the concrete canyons of city centres, seeking out the parks where it makes itself quite at home. Though it received its name in honour of a famous zoologist called Brewer, it is tempting to argue from its glazed, staring yellow eye (often seen to be bloodshot if looked at at close range through binoculars) and morning-after-the-night-before dry-mouthed song that the relationship with beer-making brewers seems more real!

Brewer's Blackbird *Euphagus cyanocephalus*

BLACKCAP

To many people's ears the beauty of the Blackcap's song approaches that of the Nightingale: indeed John Clare described it as the 'March Nightingale', a reflection both of the quality of its song and of its early arrival on European breeding grounds from winter quarters in Africa. Colloquial names include 'Mock Nightingale', 'Coal Hoodie' (from northern England) and 'Black-headed Peggy'. For the Blackcap, times are changing. Increasing numbers are now staying on over winter in western Europe instead of migrating south. This means that, although insect food is of major importance for the Blackcap, as for most warblers, through the summer months, it must look elsewhere for survival in autumn and winter. Seeds and berries are of help in autumn, but today many Blackcaps seen in winter come to garden bird tables to feed on domestic scraps of all sorts, and become possessively domineering, driving off even Starlings!

Male Blackcap *Sylvia atricapilla*

BLEEDING-HEART

The Luzon Bleeding-heart Pigeon, so called because of the tear-shaped patch of blood-red feathers in the centre of its breast, is one of a group of about a dozen 'ground doves' — only a small section of the huge (almost 200 species) and worldwide pigeon family. 'Pigeon' and 'dove' are completely interchangeable, there being no taxonomic or nomenclatural difference between the two, though in common usage 'doves' may be thought of as the softer, more placid word. The ground doves tend to come from remote islands and to be relatively small in size: the Bleeding-heart Pigeon, at about 12 inches (30 cm) and coming from (primarily) the island of Luzon in the Philippines, is no exception. Often doves are far more aggressive in display than we would like to think, and fights are not infrequent, with feathers flying. This is rather belying the Biblical connotations of doves in carrying the olive branch that is now associated with peace.

Luzon Bleeding-heart Pigeon *Gallicolumba luzonica*

BOOBY

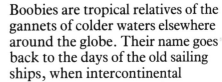

Boobies are tropical relatives of the gannets of colder waters elsewhere around the globe. Their name goes back to the days of the old sailing ships, when intercontinental voyages took months, if not years, and sailors had plentiful time in the tropics to be interested in the few other creatures they saw. Boobies proved doubly interesting: besides indicating that their nesting colonies, and thus land and perhaps fresh water, were nearby, they were also pathetically inquisitive. Hence the name, for there are numerous reports of boobies — such as the Masked Booby — perching on the shoulders of sailors in the rigging, or, less safely, landing on the gunwhale. So easily approached were they that 'stupidly' they allowed the ship's cook close enough to club them over the head, and add them, as welcome fresh meat variety, to the otherwise dreary menu of the day — usually salt pork.

Masked Booby *Sula dactylatra*

BOWERBIRD

The bowerbirds and the birds of paradise are thought to be closely related, and certainly show a close similarity in distribution, both being confined to Australasia and New Guinea or Vanuatu. Though none quite matches up to the incredible birds of paradise either for colour or for magnificence of plumes, the bowerbirds are brightly coloured and beautiful — as seems appropriate for a tropical family — in their own right. The Regent Bowerbird is one such, the male being pigeon-sized and velvety black and gold in plumage. While the male birds of paradise attract their mates by exotic displays of plumage finery, the bowerbirds rely on their considerable skills as nest-builders and decorators to achieve this goal. The bower, though, serves only to attract the female: once mating has taken place, she goes off on her own to build another nest and lay. Regent Bowerbirds build a low display platform under a bush, with an 'avenue' of two walls of upright twigs leading to it. It is decorated with leaves, berries and Lantana flowers, and parts of it are daubed with 'paint' (made of charcoal, natural pigments and saliva), the male using as a 'brush' a pad of leaves held in its beak!

Regent Bowerbird *Sericulus chrysocephalus*

BROADBILL

The broadbill family is usually placed first in the long list of families in the suborder Passeres — the perching birds — which amount to some 60 percent of the world's birds: this implies that they are a relatively primitive family, and their distribution dotted about

Africa and Asia also suggests that those alive today are but relics of a once far more important family of wider distribution. They are dumpy, thrush-sized birds, with relatively large heads and short legs, which exaggerates this stalwart impression. As their name suggests, their beaks are large, broad, flattened and strong, and much of their food is insects, either gleaned off the foliage of the wet jungles in which they live or caught, flycatcher-like, in mid-air in a sortie from a hunting perch, with occasional small lizards thrown in. Broadbills have strange circular display flights, which serve to show bright-coloured plumage otherwise concealed by the closed wings, and which are often accompanied by an extended vibrant honking cry. Broadbill nests would do credit to the weaver family so neatly are they woven, and, as an anti-predator device, the whole domed nest is suspended from a branch on a slender, twisted supporting thread.

African Broadbill *Eurylaimus capensis*

BULLFINCH

Few would argue that the Bullfinch — the male cardinal-like with scarlet breast and black cap, the female a subtle mixture of suede-browns — is a beautiful bird. But there is a 'beauty or beast' enigma, for since the Middle Ages Bullfinches have had a price on their head. In medieval England, 'one pennye' reward was offered 'for everye byrde that devoureth the blowthe of fruit', clear indication that then, as now, Bullfinches could be a seriously destructive pest because of their liking for the flower buds of fruit trees and ornamental shrubs. Their name may derive from their relatively large size (as finches go) or from their heavy- (bull-) headed appearance in flight. In Chaucer's time they were called 'Alp' or 'Alpe', which, slightly corrupted, still persists in the Norfolk colloquial name 'Blood Olp'.

Male Bullfinch *Pyrrhula pyrrhula*

BUNTING

It may well be that the Lapland Bunting gets its name from that region, where, in the brief summer months, many of the world's Lapland Buntings breed. But it may also be that this bunting's plumage, so strikingly colourful during the breeding season, has been linked with the picturesque and varied national costume of the Lapps themselves. Called the Lapland Longspur in North America (because of its long hind claw), the Lapland Bunting blends remarkably well, for all its seeming brightness, with the colours of the plants, mosses and lichens of the tundra on which it nests. Amazingly, these plumage colours are the product of wear and tear on the feathers: the Lapland Bunting wintering on western European marshes is drab, with long, warm-brown-tipped feathers. As winter progresses, the insulation is less needed, the feather fringes wear off, and the beautiful colours of the lightweight summer plumage are revealed.

Lapland Bunting *Calcarius lapponicus*

BUTCHER BIRD

Although applicable to most members of the widespread shrike family, 'Butcher Bird' is usually specifically directed at the handsome Red-backed Shrike, a summer migrant visitor to Europe that winters in Africa. Shrikes catch their prey with powerful feet and then carry them off to a perch to eat. Small items like beetles may be eaten at once, but larger prey (which includes insects, lizards and even small birds and nestlings) is often impaled on nearby sharp thorns, as if on a butcher's hook, for later consumption. Such is the impact of modern man on shrike habitat that barbed wire is now almost as prominent as thorn bushes — and the Butcher Birds have readily adapted to hanging their prey on this man-made alternative. Among a range of gruesome old colloquial names are to be found 'Worrier', 'Nine Killer', 'Throttler' and 'Destroying Angel'.

Red-backed Shrikes *Lanius collurio*

CANARY

The ancestor to the Canary, now so popular as a singing cagebird, is usually considered to be the Serin of the Atlantic Islands, a bird larger and yellower than its Mediterranean cousin of the same name. The scientific name *Serinus canarius* gives a clue both to its geographic origin in the Canary Islands and to the reason for its colloquial name. The Canary is a true finch, but has been greatly amended by its contact with man. In the wild it is grey above and rich yellow below, and frequents forests, orchards and gardens throughout the Azores. Once in captivity — it was probably first imported during the sixteenth century — aviculturists set about breeding for special colour and plumage forms, which range from purest yellow and egg-yolk orange forms to those with an untidy 'topknot' of feathers hanging down over their eyes. Often Canaries will be crossed with other finches, especially Greenfinches and Goldfinches, the progeny being called 'mules'. Interestingly, the rich bubbling continuum of song is usually as splendid in the wild as it is in captive forms, though some would argue that Roller Canaries, trained from the nest to listen to selected older birds singing, are unsurpassable for beauty of song.

Canary *Serinus canaria*

CARDINAL

The male Cardinal is one of the very few all-red birds in the world. To European eyes, perhaps bleary from a trans-Atlantic flight arriving in the USA after dark, the sight on awakening next morning of Cardinals on a garden bird table is unbelievably fabulous. Cardinals (or Cardinal Finches) are woodland seed-eaters, widespread in the southern states of the USA, which have taken readily to suburban life and to the nuts, seeds and scraps put out in feeders by bird-loving townsfolk. The female is dull brown, with rufous tinges about the tail, but distinctive because she too has a crest like the male. These are well-loved birds, and their slow but steady range expansion northwards is much appreciated by the American public. It is thought that Cardinals, which are rarely gregarious, remain paired for life. In the light of the fact that the female, unaided, does all of the nest-building and egg-incubation and undertakes most of the feeding of the young (the male may make a token contribution here), this is perhaps surprising!

Male Cardinal *Cardinalis cardinalis*

CATBIRD

The mockingbird family name is Mimidae, and they are often colloquially called 'mimic-thrushes', though in no way related to the true Northern Hemisphere thrushes like the Song Thrush, which according to Robert Browning 'sings each song twice over'. Catbirds have longish beaks, long tails and rather long strong legs, and are generally rather gawky. Although some of their relatives resemble the spotted-breasted thrushes of Europe, the Catbird's plumage is a sombre mixture of leaden greys, relieved only by a patch of chestnut beneath the tail. In other ways, too, Catbirds are something of an exception to others in their family, which in general are excellent songsters. Catbird song would be described (and then flatteringly) as a collection of brief notes and phrases, only some of which are musical! The name comes from the call, a distinctly cat-like mewing, doubly confusing as it is often produced by birds skulking deep out of sight in the bushes.

Catbird *Dumetella carolinensis*

CHANTING GOSHAWK

The goshawks are swiftly dashing birds of prey, with heavily 'fingered' rounded wings ideal for pursuing their quarry, often through dense woodland or scrub country. The chanting goshawks are African birds, named after their curious and prolonged fluting chant of a call, heard most often during the nesting season. There are two species — Pale and Dark — covering much of East Africa, overlapping but not interbreeding along the Rift Valley. Pigeon-sized, they tackle prey from the size of lizards and mantids up to birds the size of francolins. Falconers regard hawks with enthusiasm, as they take their prey by surprise after a headlong dash. It has been written of the goshawk that when 'in Yarak' — a falconers' term meaning fluffed up and with a  friendly look in her eye — she will fly with a ferocity that never seems satisfied, even with 19 kills in a day!

Dark Chanting Goshawk *Melierax metabates*

CHATTERER

The babbler family is almost 300 species strong, and is one of the most diverse assemblages of birds to be grouped in this way. Most come from the plains, scrub and jungle of southern Eurasia, and most are small or middling in size. In appearance many, including the Rufous Chatterer, seem to have been poorly put together from kits designed for several other species. The feathers are ragged and ill-kempt, the wings rather too short for their size (hence most are poor fliers) and, in reverse, both their tails and their legs and feet seem to have been designed for bigger birds. The Rufous Chatterer belongs to the largest tribe within the family, the 'song babblers', and comes from India. It is called rufous from its plumage and chatterer from its continuing chattering calls, though the latter are by no means a unique feature in this noisy squabblesome family!

Rufous Chatterer *Turdoides rubiginosus*

CISTICOLA

The cisticolas are a large group of warblers primarily to be found in Africa. Almost all are brown, some streaked, some plain, and most are small. The larger species are about 13 cm (5 inches) long, the smaller — including the Pectoral-patch Cisticola — about 9 cm (3½ inches). Most are scrub or grassland birds, and as a genus it would be hard to find their equal in difficulty of identification, so similar and so inconspicuous are they. Some experts reckon that song offers the best hopes, and a glance at field guides to African birds opens a list of amazing cisticola names, all related to their songs, and including Singing, Rattling, Trilling, Whistling, Winding, Tinkling, Wailing, Croaking and Zitting! The Pectoral-patch is a minute Wren-like bird, dumpy and with a stumpy tail, which lives on the open grasslands of East Africa and is really conspicuous only in the breeding season, when it shoots skywards at intervals and then descends, calling a monotonous 'zeet-zeet-zeet . . .' all the time.

Pectoral-patch Cisticola *Cisticola brunnescens*

COCKATOO

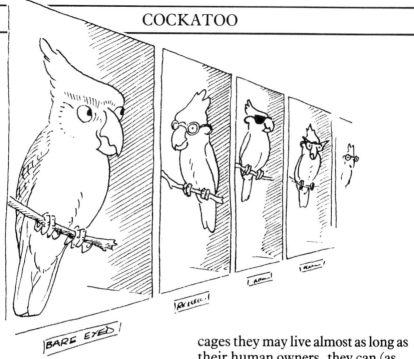

White and with lemon-yellow crests, the members of the cockatoo family — including the Bare-eyed, familiarly known as the Little Corella — are very much part of the Australian scene, as raucously noisy in the bush as a Sydney bar in mid-evening. Cockatoos, like their relatives in the parrot family, have long been kept in captivity and as pets have an amazing capacity for mimicking human speech. As in cages they may live almost as long as their human owners, they can (as many zookeepers know) acquire a vocabulary of some considerable magnitude and scope. Though in homes and zoos much of what they mimic comes from children, many cockatoos have at some time been pets of seafaring men. Their vocabulary ranges into swear words, and worse, well beyond the familiar 'hello' and 'who is a pretty boy then?'!

Bare-eyed Cockatoo *Cacatua sanguinea*

COCK-OF-THE-ROCK

Cocks-of-the-rock — the Andean is one of the two species — are exotic, almost legendary birds of the Amazonian jungles of South America. Their lurid and bizarre black and orange plumage, with a helmet-like crest almost concealing the beak, has been known since the earliest explorers brought back specimens to exhibit before unbelieving European eyes. Less has been known of their display, which, because of the jungle gloom, has only recently been well filmed. The males gather at a collective 'lek' or dancing arena, with calls strikingly similar to human wolf-whistles. One by one they drop from the branches, like untidy orange dusters, to their chosen patch on the forest floor. There they dance, with movements as formless as modern human dances (but much less energetic), until a watching female makes her choice by tapping one on the back. Mating takes place immediately, and she vanishes back into the jungle to lay her eggs and raise her young as a one-parent family, the male taking no further interest.

Male Andean Cock-of-the-rock *Rupicola peruviana*

COOT

The term 'as bald as a Coot' has obvious connections with the white waxy patch of skin on the Coot's forehead, but it is more difficult to discern why this waterbird should be the butt of terms like 'as crazy as a Coot'. Certainly, whatever their problems, Coots are successful birds, occurring in flocks during the winter months on lakes and reservoirs deep enough to allow them to dive for food. So buoyant are they that diving seems to present some difficulties, and Coots often make a little jump and splash out of the water before plunging down in search of pondweed. They bob up just as buoyantly, sometimes stern-first! In the breeding season, pairs of Coots still prefer larger reed-fringed fresh waters, and are fiercely territorial. It may be these sudden unprovoked wings-raised, feet-kicking attacks on the neighbours that earn the name 'crazy'.

Coot *Fulica atra*

CORDON-BLEU

Among the smallest of the African waxbills, the cordon-bleus are easily recognised by their azure-blue head, breast and tail, some species having a dark brown crown, others a blood-red patch on the nape, or on the cheek like the dainty Red-cheeked Cordon-bleu. Tit-sized, the cordon-bleus are too small in themselves ever to have become a gastronomic delicacy except to the most hungry of bushmen, and, as they rarely occur in flocks, do not even make up in quantity what they may lack in quality. Though primarily birds of the arid grasslands and scrub of eastern Africa, feeding much of the time on small seeds on the ground, cordon-bleus have a liking for over-ripe fruit. To such a delicacy they can easily be tempted, appearing as a sparkling flight of indigo, from which, doubtless, their name is derived.

Male Red-cheeked Cordon-bleu
Uraeginthus bengalus

CORMORANT

As its name implies, the King Cormorant is one of the larger members of its family, breeding on the Falkland Islands and on the adjacent coast of Tierra del Fuego. An alternative name is Rock Cormorant, which aptly describes the sort of coast on which huge cormorant colonies form. This seems often to be the case with Southern Hemisphere cormorants, nowhere better exemplified than by the huge Guanay Cormorant colonies on islands off the Peruvian coast. King Cormorant pairs tend to return to the same nest season after season, and the seaweed, flotsam and jetsam accumulates and accretes into a concrete-hard mound reminiscent of a giant flamingo nest. Throughout incubation, the male brings home 'gifts' of seaweed with which the female further adorns the nest: burglary is rife, and should she turn her back a neighbour is likely to lean over and remove it. Like other cormorants, and despite the presence of the family through more than 30 million years of evolution, King Cormorant feathers are not fully waterproof, and birds must occasionally stand out to dry off, lest they become waterlogged and sink!

King Cormorant *Phalacrocorax atriceps*

CORNCRAKE

Though often frequenting fairly marshy meadows, the Corncrake is the most land-living of the crake (or rail) family. In fact, it is often known as the Land Rail. It is now restricted largely to northwest Europe, where the farming pattern is such that the hayfields are late to be harvested, and often cut by hand. Elsewhere, improved farming techniques and modern machinery allow hay to be cut much earlier in the season, and prevent the Corncrake (a late-arriving migrant from Africa) from breeding successfully. Its scientific name is *Crex crex*, which is a fair rendering of its monotonous call, often produced with ventriloquial effect and at all hours of the day and night. Corncrake numbers are falling fast, and there are worries over its impending extinction. Despite this, in areas where the practice of growing grass for silage is increasing and where Corncrakes have attempted to make a come-back in the new habitat, the local human population has promptly complained at being kept awake all night by their raucous calling.

Corncrake *Crex crex*

COWBIRD

Diversity in social organisation could be said to be a characteristic of the large Icterid (or American oriole) family. Within the family can be found birds that are highly gregarious at all seasons and breed in colonies, through those breeding in only loosely-knit colonies, to species that are quite solitary. In addition, the family contains a group — the cowbirds — which are among the handful of avian groups to have developed nest parasitism. Nest parasitism is almost the rule among cowbirds, the majority of which are polygamous, as most make no nest of their own but normally lay into the nests of 'foster parents'. As a forerunner of this habit, the monogamous Bay-winged Cowbird is not a parasite, but does take over disused nests of other species, very rarely building its own. Most of the remaining cowbirds, like the Brown-headed, are not as choosy as the Eurasian Cuckoo when it comes to 'hosts' and will lay in a variety of nests, some of related species. Female cowbirds keep a watch on likely foster parents, not laying until the foster clutch is under way. Unlike Cuckoos, the legitimate inhabitants of the nest often remain and fledge successfully with the parasite.

Brown-headed Cowbird *Molothrus ater*

CRANE

The display dance of the Crowned Crane is surely one of the most spectacular of African wildlife sights. Cranes are large birds, more than a metre (39 inches) tall and with a 2-metre (6½-foot) wingspan, and all indulge in leaping and bounding displays, wings outstretched, trumpeting and honking weirdly, in their attempts to secure a mate. So striking is the plumage of the Crowned Crane, especially the huge white wing patches, that this spectacle is visible from a great distance and may

attract predators. Even lions have been known to approach the huge nest, built of grasses on the ground out on the open plains, with evil intent, but the cranes, whooping wildly, golden crests bristling and staring eyes glinting fiercely, mount such an effective show of aggression that the would-be predator departs to find easier prey.

Crowned Crane *Balearica pavonica*

CROSSBILL

Like miniature parrots, Crossbills clamber about with some ease but very little grace around clusters of cones at the ends of conifer branches. Their heavy, hooked beaks give them a top-heavy look, and rather ponderous movements heighten the similarity to the parrot family. The two halves of the beak are twisted, the tips overlapping like scissors, the whole providing the ideal tool to extract the conifer seed and its nutritious kernel from within the woody protective cone, which is shredded and dropped to the ground below. A striking experience of this nature may be all that alerts the unwary birdwatcher to their presence high above in the treetops. Pine cones may mature during the winter months, and Crossbills will capitalise on the food source and start breeding. Such a strategem, though effective on a nutritional front, leads to other problems, as in their northern conifer-forest homes incubating Crossbills must sometimes sit with snow on their backs.

Crossbills *Loxia curvirostra*

CROW

Hooded Crow *Corvus corone cornix*

Crows in general, partly because of their sombre plumage, partly because of their predatory and scavenging feeding tactics, and partly because of the views gamekeepers have of them, are even today regarded as evil birds. The origins of such beliefs go back to pagan times, even though St Cuthbert — as one of the earliest saints with bird connections — was befriended by crows on the Farne Islands. Interestingly, it is thought that *crow*bars, with their powerful prising capability, are named after the strong hooked beak of the bird, and the Latin scientific name for the crows — *Corvus* — is the Roman name for a grappling hook on their galleys. The hood of the Hooded or Grey Crow is a conspicuous feature, but one variable in shape, as hybrids are frequent between Hooded and all-black Carrion Crows where their ranges overlap.

CUT-THROAT

The Cut-throat belongs to a large family — the weaver-finches — which inhabit most countries in the warmer areas of the Old World. Many of them are best known as cagebirds, because as small seed-eaters they are relatively easy to keep and because many are brightly coloured. As with all captive creatures, they are to be seen and enjoyed far better free and in the wild. The Cut-throat belongs to a group called the mannikins, which are mostly rather dull-plumaged jobs. The name derives from a blood-red 'gash' of feathers extending from ear to ear under the chin, but beyond that it is not a bloodthirsty bird. Its only bad habit is that common to many creatures including humans — it is lazy. Rather than build their own nests, Cut-throats take over those of others, such as the pendant snake-proof nests of some sunbirds, occasionally before the rightful occupants have finished raising their young!

Cut-throat *Amadina fasciata*

DABCHICK

The Dabchick or Little Grebe is the smallest of the grebe family. A neat waterbird, it spends most time on or under water. Its lobed feet are positioned right back near its bristly, almost non-existent, tail — where they give most effective propulsion. Dabchicks build floating nests of waterweed, sensibly tethering them to nearby emergent reeds, and these nests rise and fall with the water level, rather than being inundated or left embarrassingly high and dry as those of Moorhens so often are. They are poor fliers, requiring an extended pattering take-off run to get airborne, and usually dive rather than fly for safety. As cautious birds, they rise slowly to the surface until just the head emerges, often draped in weed, like a miniature Neptune. If danger continues to threaten, the Dabchick sinks slowly from sight like a tiny crocodile, leaving just its beak and nostrils showing.

Dabchick *Tachybaptus ruficollis*

DIPPER

Birds of sparkling streams, flowing over shallow, pebbly rapids and with a plentiful supply of large boulders for perches, Dippers are birds of hilly or mountain regions. Thrush-sized but Wren-shaped, they perch on boulders in mid-stream, bobbing up and down all the time. Their feet seem several sizes too large, but with good reason because the Dipper has a unique feeding technique. It slips off its boulder into the passing torrent with hardly a ripple, sometimes wading waist-deep in the shallows, sometimes submerging completely. Beneath the surface it seeks small fish, caddis fly larvae and shrimps as food: often it will walk along the bottom, using those powerful feet to resist the pull of the current and its own buoyancy as it hunts. In deeper water it may also swim, using its wings for propulsion, appearing silver in a thin sheathing air-bubble, before bobbing to the surface and either returning to its perch or whirring away, low over the water on stubby wings, to another part of its territory on the river.

Dipper *Cinclus cinclus*

DOVE

Doves — and pigeons, there being no distinction between the two save the difference in name — range in size around the world from the lark-sized Namaqua Dove to the turkey-sized crowned pigeons. The Tambourine Dove, flying as swift and straight as an arrow, is an African bird, favouring open woodland or thick bush. It feeds on the ground on various seeds, so is not a true fruit-eating forest pigeon, but does nest in trees. The nest is flimsy to the point of transparency, a collection of twigs placed untidily over a fork, so that often the round white eggs can be seen from below as one looks up *through* the nest floor. This shaky structure supports usually two beefy youngsters, which are fed 'pigeon's milk' by their parents. This 'milk' is a greyish fluid of great nutritional value secreted by the walls of the crop, a feeding technique unique to the pigeons and doves.

Male Tambourine Dove *Turtur tympanistria*

DOWITCHER

Most waders are tundra birds, exploiting the richness of the Arctic summer, and thus unlike most other birds probably reached their peak while the world was in the grip of the Ice Ages and the tundra zone was at its greatest. In the mild period before the Ice Ages, most waders would have been confined to tundra 'refuges' in Greenland and Siberia, and it is from the latter that the dowitchers spread into the Americas, where there are two species, Long-billed and Short-billed (though the differences are by no means as extreme as the names suggest). They are the only snipe-like birds to be seen out on the open mudflats. Here they probe for worms and shellfish, with a fast-moving action that has been likened to a sewing-machine needle. The end of the beak (besides being well equipped with sensory nerves assisting prey location and identification) is also flexible, and just the tip can be opened by special muscles on the skull, a feat made possible because the nasal bone — the 'bridge' of the beak — is not fixed rigidly to the 'face': it can be pulled up onto the 'forehead', answering the question 'Why do they not get a mouthful of mud each time?'

Short-billed Dowitcher *Limnodromus griseus*

EAGLE

Relatively small (roughly Buzzard-sized), the Booted Eagle occurs in two colour phases when adult. Most of the larger birds of prey have a number of immature plumages, changing as they mature, but this just adds to the difficulties of identification in an already notorious group. These are birds of wooded hillsides, venturing out infrequently onto nearby plains and mountain slopes in search of their food, which is mainly small mammals and birds. The Booted Eagle is a real forest bird, hunting often beneath the canopy and nesting as often in a tree as on a rocky crag. Although, with narrower wings, it is more agile in the air than a Buzzard, it tends to take most prey on the ground, relying on the stunning force of the impact as it plummets down, and the cruel crushing and stabbing power of its spreadeagled toes and inch-long (2½-cm) talons to inflict a mortal wound. Distributed widely from Turkestan to Spain, in some Mediterranean countries the Booted Eagle is known — and hated — by the local human inhabitants as the 'chicken hawk' for its love of the easy living afforded by domestic poultry.

Booted Eagle *Hieraaetus pennatus*

FALCON

Although called a falcon, the Laughing Falcon is much rounder-winged than most of its relatives, and is placed in a separate group. It inhabits widely varying countryside in tropical America, ranging from cactus-studded deserts through to the dense rain forest of the Amazon, but in each habitat the major food of the Laughing Falcon is the same: snakes. The risks involved in such a diet are considerable, so Laughing is perhaps hardly the best name. Included in the menu are the highly poisonous coral snake and its equally colourful but harmless imitator, the false coral snake. The falcon's hunting technique is to drop down on its prey, usually achieving a kill almost instantly with needle-sharp talons. Often the snake is decapitated before it is carried away, parallel to the body like a torpedo, but its tail trailing out behind like the tail of a child's

kite. Noisy birds, especially when they call in duet in early morning or evening, Laughing Falcons get their name from their rather hysterically (understandably so) shrill peals of laughing calls.

Laughing Falcon *Herpetotheres cachinnans*

FIRECREST

The Firecrest has the distinction of being one of the smallest, if not *the* smallest, European bird, weighing in at about 5 grams or half-a-dozen Firecrests to the old-fashioned ounce! Despite their diminutive size, Firecrests are migrants, capable of crossing the North Sea at times in autumn, when the weather can be cold and rough. Amazing that the appropriate navigational ability can be housed just in the head of a bird so tiny. The fiery crest is flared open to reveal its flame-coloured heart during bouts of aggression and display. The Firecrest, for all its tininess, is a fiery customer, reacting strongly to intruders into its territory be they of its own kind or the related Goldcrest. Even a portable cassette recorder playing a tape of Firecrest song will incite the territory-holder to such a degree that he may attack the machine and dance in fury on the loudspeaker.

Firecrest *Regulus ignicapillus*

FLYCATCHER

Despite a dowdy plumage, an almost total lack of song and a generally inconspicuous way of life, Spotted Flycatchers remain firm favourites throughout the temperate climatic regions of Europe. Originally birds of open forest glades, Spotted Flycatchers are among the birds that have adapted well to, and often thoroughly exploited, man's gardens. Buildings and plants provide an abundance of nest sites: these flycatchers prefer a projection giving a clear view but offering shelter from the sides and overhead — a creeper or vine on a house wall is thus ideal, as is an open-fronted nestbox, and such sites are used year after year. When hunting insects (not just flies but all sorts and sizes — Spotted Flycatchers can become involved in undignified wrestling matches with heavy-bodied hawk moths), they choose a prominent vantage point, dash forth to snap up a captive, then return to the same spot to consume their meal.

Spotted Flycatcher *Muscicapa striata*

FRIARBIRD

The Noisy Friarbird is one of the Australian 'wattlebirds', a group of the honeyeaters that have areas of bare skin or wattle-like flaps about the head. Jay-sized, the Noisy Friarbird is one of the bulkier of the honeyeaters, and one of the more raucously vocal — hence the 'Noisy' — as it moves about scrub, open bush and suburban garden, often going on the rampage among the fruit and vegetables. Though their beaks are general-purpose enough to tackle fruit, the tongue is specially adapted to sipping nectar, with sides rolled up and a brush-like tip. Many Australasian trees and shrubs are pollinated by visiting honeyeaters. The drab plumage colours of the friarbird, and the untidy ruff of coarse feathers about its neck, suggest the caped habit of a monk or friar, and this similarity is heightened by the bird's (rather repulsive) bald pate which resembles a friar's almost-bald haircut or tonsure. Australians lacking a desire to indulge in ecclesiastical niceties simply and rudely refer to the bird as 'leatherhead'!

Noisy Friarbird *Philemon corniculatus*

FRIGATEBIRD

Called 'man-o-war birds' by mariners in the days of sailing ships because of their piratical treatment of other seabirds, frigatebirds occur in most tropical seas. They have an enormous swallow-tail and narrow, angular wings with a huge span, often in excess of 3 metres (10 feet), giving excellent control and making them among the most expert fliers. Their flight skills and long hooked beaks are thought to have evolved originally to catch flying-fish which indulge in brief sorties above the waves. Now they are put to other uses, ranging from in-flight 'smash-and-grab' raids, when they steal nest material from other birds' nests, to (again in flight) preying on eggs and nestlings, which they snatch from an unguarded nest in passing. Out at sea, they tend to harass other seabirds flying back to their nests with food for their chicks, forcing them to disgorge and catching the food before it hits the sea.

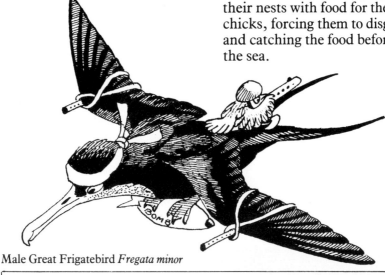

Male Great Frigatebird *Fregata minor*

FROGMOUTH

The frogmouths are a peculiar family, allied to the nightjars, occurring in Australasia and Malaysia. They share the magnificent camouflage plumage of the nightjars, and tend to spend the daylight hours perched, some horizontal on a branch but others, such as the Tawny Frogmouth, bolt upright at the angle of a branch, where they look just like a broken-off limb. Large eyes — for nocturnal hunting — are closed during the day to protect them from the sun and to improve concealment. Their song is a foghorn-like 'oom-oom', not unlike the call of a bullfrog. They nest in trees, not on the ground like nightjars, using their own feathers to help make a shallow nest with twigs, on which a couple of owl-like white eggs are laid, quite lacking the superb speckled camouflage of eggs of related species. Though they have the huge frog-like mouth, opening from 'ear to ear' as in the nightjars, frogmouths tend to hunt more like owls, dropping onto prey passing beneath their perch. Much of this is beetles, scorpions and worms, but Tawny Frogmouths will catch unwary mice, and, ironically, frogs!

Tawny Frogmouth *Podargus strigoides*

GO-AWAY BIRD

Go-away birds are large, drab, grey and white relatives of the colourful — often splendidly so — turacos. With a chicken-sized body and a banded tail of Magpie-like proportions, they seem ill-fitted for their largely arboreal life in the thorn bushes and acacia trees studding the dry East African grassy plains. Their diet is composed largely of fruits and seeds, though the occasional insect does not come amiss, and the White-bellied Go-away Bird and others in the family are often colloquially called 'plantain eaters'. They are prone to stretch out for a tempting morsel of food just beyond reach, but, lacking the agility and grace of families like the tits, tend to end up (literally) in undignified upside-down postures, clawing and flapping frantically to maintain a grip. Their name comes from their harsh 'go-away aaah' call, which is intermingled with rather hysterical chuckles.

White-bellied Go-away Bird *Corythaixoides leucogaster*

GOOSE

Ne-nes, or Hawaiian Geese, are neat middling-sized geese with brown and white plumage. Their feet are only partially webbed, looking for all the world as if the webs had been worn away on the jagged larva rocks of their home islands in Hawaii. The Ne-ne was recently threatened with extinction by rats, which invaded its nesting islands from visiting ships. Disaster was staring the species in the face when a rescue operation, masterminded by Sir Peter Scott, brought a handful of survivors to European waterfowl collections. The Ne-ne took to captivity like the proverbial duck (or goose) to water, rapidly becoming tame and soon pestering visitors for titbits of food to augment its well-thought-out and already luxurious diet. Despite the tiny original numbers, a carefully planned and monitored breeding programme has been a great success, and many Ne-nes have now been re-introduced to predator-free islands in their homeland.

Ne-ne *Branta sandvicensis*

GRACKLE

The Great-tailed Grackle is one of about 100 species of Icterids. This family, all possessed of dagger-like beaks of varying lengths, is peculiar to the New World, with most species occurring in the tropics. Confusingly for Europeans, many of them are referred to as 'blackbirds', and the Great-tailed Grackle is one of these. As the male is often more than a foot (30 cm) long, it would dwarf its unrelated European counterpart. About half of this length is the grotesquely large tail. The female is brown and considerably smaller than the yellow-eyed glossy-black male, whose plumage shows a purple iridescence in the sunlight. The tail is not only long but large, and shaped like a cupped fan, or the keel of a boat — its very similar relative is known as the Boat-tailed Grackle. It is often twisted from side to side as well as wagged in slow motion. In flight (which is anyway not of the strongest) the tail seems on occasion to take over and dominate the movements and stability of the bird to which it is attached, especially in strong winds.

Male Great-tailed Grackle *Quiscalus mexicanus*

GRENADIER

Even a guardsman as elegantly uniformed as a Grenadier in ceremonial dress would blanch at the plumage colours of his avian namesake, with its russet head and shoulders, bright cobalt-blue eye patch and underparts, and bright red beak. The sealing-wax-red beak gives the clue to the Purple Grenadier's family — seed-eaters called 'waxbills'. These birds, mostly sparrow-sized or smaller, are commonest and most widespread in Africa. They usually form stable pairs, sharing the tasks of nest-building, incubation and care of the young, but despite this they are social, gathering in groups and indulging in communal preening and 'billing-and-cooing' sessions — all of which tends to make them popular as cage and aviary birds. In the aviary, the elaborate and (to human eyes) quaint courtship dance of the male can be studied: this seems mostly to involve a series of stiff-legged leaps up from the perch, followed by a rather mesmeric swinging from side to side while standing in one spot, showing the female the most colourful plumage areas to best effect.

Purple Grenadier *Uraeginthus ianthinogaster*

GUINEA-FOWL

Helmeted Guinea-fowl *Numida meleagris*

Helmeted Guinea-fowls, as all guinea-fowls, are among the tastiest of African gamebirds, and among the seemingly most stupid. The tiny size of their head (even with its grotesque waxy, fleshy head ornamentation) relative to their bulky, droopy-apron-girt body gives some indication of their likely brain power. 'Seemingly' because, despite their popularity for the table or cooking pot, whether secured by gun, bow-and-arrow or snare, they remain common and widespread through open bush country and derelict farmland. Guinea-fowls are largely terrestrial feeders, like giant chickens, tending to feed in large colourful flocks. If danger threatens, the whole flock panics readily in a turmoil matched by no other bird, some dashing off, others trying to take wing and crashing into nearby bushes in a confusion of spotted feathers.

GULL

Sooty Gull *Larus hemprichii*

Most of the world's gulls have plumages predominantly white but involving also blacks and greys. The Sooty Gull is one of a tiny select group of the middling-large gulls which have largely grey plumage, with a near-black head and very little white, mostly on the belly. The Lava Gull of the Galapagos and the Red Sea Black-headed Gull are similar. An alternative name for the Sooty Gull is the Aden Gull, which gives an indication of its range — from the shores of Arabia south down the east coast of Africa to Zanzibar. Sooty Gulls are scavengers, and, like their temperate-latitude cousins the Herring Gulls, successful. Some of their scavenging is at refuse tips or village middens, but more often they are to be found on beaches where fish is being gutted or at the village markets — often impromptu ones — wherever fish is being sold. Here they show little fear of man as they dart about after scraps. An examination of the bones in their pellets, or castings, naturally indicates an artificially wide diet, as so much of what they eat is procured by man, but sometimes even stranger artefacts turn up, such as the limbs of plastic dolls!

HAMMERHEAD

Hammerheads — or Hammerkops to give them their Afrikaans-derived local name — are among the smaller and drabber members of the herons living in Africa, a country rich in wetland birds. They feed on fish, frogs and shellfish by small muddy streams and pools, and are so well camouflaged in this habitat that to discover one close by after several minutes of birdwatching can come as a shock. The crest of head feathers that gives rise to the name — in fact pickaxe would be better than hammer — does not serve as a counterpoise to the beak but functions primarily in display. The most remarkable feature of the Hammerhead is its nest. It is huge, built of mud and branches, and set high in the trees. It is a hollow structure, with a side entrance, often several feet in diameter, and the whole may weigh half a ton (500 kg)!

Hammerhead *Scopus umbretta*

HANG-NEST

The diversity of nest-building capability that can be seen in just one family is nowhere better exemplified than in the great American oriole family, the Icterids. Some are nest parasites, making no nest at all, others make simple cup-like structures on the ground or in bushes. Most advanced are the various suspended nests. The North American 'blackbirds' often breed in marshy surrounds, and hang their cup-like nests onto the reed stems over the water and out of reach of predators; but, in the tropics, the hang-nests — oropendolas and caciques — such as the Brazilian Troupial build masterpieces. The birds themselves are colourful and somewhat Rook-like in size and appearance. They are colonial (and usually polygamous) and weave their nests beautifully from strips of grass. Often 100 or more nests in a colony will hang from the topmost branches of tall forest trees, and some may be 2 metres (6 feet or more) long. They look for all the world like an array of giant Christmas stockings, hanging up and waiting for gifts to be inserted. The entrance slit is at the top, and in the pouch at the base eggs or young rest safe from predators.

Brazilian Hang-nest *Icterus icterus jamaicaii*

HERON

Goliath Heron *Ardea goliath*

As they often stand getting on for man-high, it is no wonder that Goliath Herons were named after that massive Biblical figure. Their length of leg allows them to wade deep out into the African swamps, where among the water lilies they pause, motionless, often for some minutes, before securing their prey with a sudden stab. Most popular among their prey are fish, though giant bullfrogs and even unwary waterbirds and mammals will be taken if they stray too close. The robust beak is the length of a man's forearm, and 46-cm (18-inch) fish — catfish and black-mouthed bass are favoured — are often taken. A fish of this size may be a little difficult to subdue and swallow (head first, or the fins get stuck in the heron's throat), especially if the aim has been a shade off target and both mandibles have pierced the fish, effectively clamping the beak shut!

HOBBY

The Hobby is a summer visitor to the warmer parts of Europe, wintering in Africa. It is one of the most elegant and probably the fastest-flying of the falcons, its long sickle-shaped wings giving it the power and speed, in a sprint, to overhaul and catch a Swift. Though timing flying birds is difficult, Hobbies must often exceed 60 mph (95 kph) and may, on occasion, reach the magic 100 mph (160 kph) that is sometimes attributed to them. But not all a Hobby's hunting is necessarily so dashing: they often catch large insects such as grasshoppers and dragonflies, and will even eat them clutched in one foot in mid-air while patrolling their hunting grounds. It is said that their name is derived from an old French verb *hober*, meaning 'to stir', and this certainly is a fair description of their most commonly used hunting technique. Choosing an area such as a weedy field with a large finch flock enjoying the seeds, or a lake with many Swallows and martins taking advantage of the abundant insect life, the Hobby will suddenly dash in among the birds, stirring up terrified confusion, during which it grabs its prey.

Hobby *Falco subbuteo*

HONEYEATER

The Spiny-cheeked Honeyeater, the size of a large thrush, is one of the most widely distributed members of this Australian family, occurring in open bush and grassland over most of the country, and really absent only from the extreme north and from the forests of the east coast. Although occasionally taking fruit or insects, Spiny-cheeked Honeyeaters, as their name implies, favour feeding on the sugar-rich nectar of eucalyptus and other flowers. In consequence much of their life is spent in the trees, where, with a longish beak and flexible neck, they can with little difficulty reach into sprays of flowers to remove the nectar. This they do with a tongue that in the course of evolution has become very different from that of other birds. It is very long, and can

be stuck out to a degree that any ill-mannered child would envy. The sides roll up to give two long grooves and the tip is incised deeply to give it a frayed appearance. This combination of brush and drinking straw is highly effective in transferring the energy-rich food from plant to honeyeater.

Spiny-cheeked Honeyeater *Anthochaera rufogularis*

HONEYGUIDE

The drab honeyguides of Africa have evolved the most extraordinary co-operative feeding technique. They love the honey, and particularly the grubs, from the numerous nests of wild bees to be found in the tropical forest trees and caves. Despite their plain plumage — as evidenced by the Thick-billed Honeyguide — and lack of any conspicuous song, they are reputed to be able to attract the attention of other predators on bees' nests, such as the ratel (honey-badger), even some way away from a potential food source, and then to lead them to it. Once the ratel has demolished the nest, the honeyguide can eat its fill. Legend has it that African natives will serve the same 'major predator' role, but that, if they do not leave adequate supplies for their guide, dreadful retribution will follow when a honeyguide leads them into the path of a hunting leopard!

Thick-billed Honeyguide *Indicator minor conirostris*

HOOPOE

Hoopoe *Upupa epops*

Strikingly plumaged in sandy-pink and boldly barred black and white, Hoopoes are conspicuous birds. Their long crest may be raised in annoyance, or during display, and seems always to fan out (as if under the influence of gravity or deceleration forces) whenever the bird lands. It is not just the plumage that is conspicuous: their 'pooh-pooh-pooh' call, though not really loud, is far-carrying, and following a calling bird through the olive groves may lead to the discovery of a nest to which the word 'pooh' could also be applied. Hoopoes are hole-nesters, feeding their young on insects and a large number of lizards and the like. The young seem to have little sense of nest hygiene (and, as many other birds, no sense of smell) and sit surrounded by flesh rapidly decaying in the heat of the sun. Following such a 'scent gradient' is an excellent way of locating the nest of the otherwise secretive Hoopoe.

HORNBILL

However huge, the beak of the hornbill is a lightweight structure, its size assumed to be related to its function in display. Despite an apparent clumsiness, it can be used with delicacy to handle small fruits and seeds, which are tossed back into the throat. Surprisingly (because of their size), hornbills, of which the Pied Hornbill is a typical example, are hole-nesters, but with a difference! Once the female begins incubating after laying her first egg, the male cements up the hole with his mate inside, leaving a tiny slit through which he passes food. This effectively excludes predators. After the eggs hatch, the female hacks her way out to help feed the young. Amazingly, the young then cement up the hole with mud and droppings behind her, until the first chick is ready to leave. This chips its way out, and the remaining chick again cements up the hole for protection until it, too, is ready to emerge and face the outside world.

Pied Hornbill *Anthracoceros malabaricus*

HUMMINGBIRD

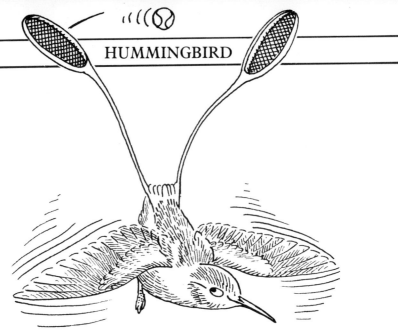

The flight prowess of hummingbirds often seems to extend beyond the bounds set by the physiological capabilities of their muscles, lungs, heart and blood supply. Their wings beat so fast that to human eyes they are just a blur, and in so doing make the humming noise that gives the bird its name. Even the Racquet-tailed Hummingbird, its blob-ended tail feathers extraordinarily elongated for display, can fly forwards, backwards, up and down as it hovers in front of flowers, sucking out energy-rich nectar with its specially adapted tubular beak and tongue. Female hummingbirds, however, such as the female Booted Racquet-tail in the photograph, lack the ornamental tail plumes possessed by many males in this family of many species. Special joints and muscle layouts allow the wings to beat over a 'figure-of-eight' configuration, which gives the hummingbirds of the Americas their fantastic flying dexterity.

Female Booted Racquet-tail *Ocreatus underwoodii*

IBIS

Ibises are long-legged wading birds of tropical and semi-tropical wetlands. They probe deep in the mud for worms and shellfish — and the occasional sleepy fish — with their long downcurved beaks. With its black and white plumage and lacy white filamentous back feathers, the Sacred Ibis should be an elegant bird, but the effect is often marred by the drab or soiled nature of the plumage, which comes from the ibises' habit of also feeding on refuse and carrion on generally unsalubrious and muddy refuse tips. Held in high esteem (not that it did them much good!) by the ancient Egyptians, Sacred Ibises acquired their name because tens of thousands of them were captured, killed and embalmed — as avian mummies — and incarcerated in tombs like the pyramids. The numbers of ibises despatched to their happy hunting grounds rose according to rank, the largest numbers being entombed with the Pharoahs.

Sacred Ibis *Threskiornis aethiopicus*

JAY

Compared with the birds of tropical regions, few European birds are brightly enough coloured to be called gaudy, certainly not many year-round residents. The Jay is one exception. Their harsh penetrating cries and bright colours are ideal for Jays to keep in contact with their fellows in the dense cover of their woodland habitat. Jays are among the birds that hoard food in autumn for consumption later in the winter when times are hard. Acorns are their favourite, and the end months of the year see Jays flopping awkwardly across the fields, acorn in mouth. How many of these acorns that they bury they find again is open to argument: one view has it that their inability to find most has been the origin of many new oakwoods!

Jay-walking — crossing the road carelessly — is of American origin, and may not refer to Jays on the ground but may be a corruption of the phrase Jay-hawker, referring to guerrilla bands in Texas, last century, making sudden destructive attacks in much the same way as Jays prey on other birds' nests.

Jay *Garrulus glandarius*

KILLDEER

The Killdeer is a large American member of the great plover family. With its double black breast bands and loud 'kill-deer' call, it is one of the most conspicuous of waders. More so, perhaps, because it has adapted well to man-made habitats such as airfields, farmland and extensive lawns, and even gravelly roadside verges as additions to its typical beach nesting areas. The nest is little more than a shallow scrape, containing a clutch of four excellently camouflaged eggs. Despite this natural concealment, the Killdeer, like several other plovers, indulges in an elaborate 'distraction display' whenever a potential predator ventures near the nest. Colloquially this ploy is known as the 'broken-wing trick'. The adult bird runs close in front of the predator, trailing one wing along the ground, sometimes stumbling, sometimes falling over, but always leading the predator away from the nest with the tempting 'carrot' of an easy meal just ahead of it. Should the predator get *too* close, then the Killdeer jumps noisily, and effectively, into a well-timed escape flight!

Killdeer *Charadrius vociferus*

KINGFISHER

Often all that we see of a Kingfisher is an electric-blue arrow, shooting past on whirring wings low over the water, jinking at the last minute to avoid overhanging branches. According to the ancients, the Kingfisher — called 'Halcyon' — made a nest of fish bones and launched it on the sea. While she was brooding, the gods ordered that the oceans be calmed. Pliny, normally an acute observer of natural history, wrote (at about the time of the birth of Christ) 'they breed in winter, at a season called *the halcyon days*, wherein the sea is calm and fit for navigation'. In actuality, the nest is anything but halcyon. Kingfishers breed during the summer, excavating a tunnel up to a yard long in a suitable bank. In a chamber at the end of this lies the nest — made of fish bones yes, but dark, slimy, noisy and above all smelling powerfully of aged fish and the droppings of the young. So revolting is it that the parents often dunk themselves in the water immediately after leaving the tunnel to remove the scales and slime.

Kingfisher *Alcedo atthis*

KNOT

The delightful russet summer plumage of the Knot, which blends so well as camouflage with the mosses of its Arctic tundra breeding grounds, is rarely seen to best effect in western Europe, as Knots pause only briefly to 'refuel' on estuaries before hastening northwards in spring. More familiar is the Knot as a wintering bird, and flocks (occasionally hundreds of thousands strong) frequent favoured sheltered bays and estuaries from Britain to far south in Africa. In flight, as they twist and turn (miraculously with no collisions), these huge flocks of dull grey birds look like smoke billowing in the wind. On the mudflats, Knots congregate in dense flocks, standing and feeding shoulder to shoulder — hence the name. Often they feed at the edge of the incoming tide, and hence the scientific name *canutus*, after the old British king Canute who in legend attempted to use his kingly power to prevent the tide coming in.

Knots *Calidris canutus*

LAMMERGEYER

The Lammergeyer is an aberrant vulture, as its alternative colloquial name Bearded Vulture implies. Though widely distributed on a world scale, nowhere are Lammergeyers frequently seen, which makes good views a major achievement for any birdwatcher. They are birds of remote mountain areas, often parched, arid country, and nest on crags. Although they have a huge wingspan, their wings are much more slender than those of most of their cousins, and they have a distinctively long diamond-shaped tail. They are carrion feeders, but have developed a special technique that allows them to exploit parts of the carcase that other vultures cannot reach. Lammergeyers may take a bone — normally a large one — in their beak and ascend to some height before dropping it: on impact the bone breaks and the Lammergeyer removes the marrow with its specially scoop-shaped tongue. Even human skulls are on record as having been subjected to this grisly treatment.

Lammergeyer *Gypaetus barbatus*

LAPWING

Lapwings are a group of waders — part of the large plover family — of worldwide distribution, particularly noteworthy for their short, rounded wings. This gives them a floppy flight, often enhanced by black-and-white wing feathers, hence 'lap-wing'. The Long-toed Lapwing is one of several African members of the group. It is conspicuous in its length of leg and particularly length of toe, because it breeds and feeds near freshwater swamps, often walking, lily-trotter-like, out over the floating leaves, where its feet prevent it sinking. Like other lapwings, it puts its wings to good use as a 'terror' weapon in defence of its eggs or young. The rounded wings with their fingered ends produce an impressive and frightening noise when the adults dive-bomb intruders, certainly capable of distracting a predator's attention. Often the 'intruder' is a harmless bystander, and many a partridge or francolin, or even goat or sheep, has had to duck and scurry for cover in an undignified way to avoid the aerial wrath of the lapwing.

Long-toed Lapwing *Vanellus crassirostris*

LAUGHING THRUSH

The laughing thrushes — the White-crested is one of about 150 species — comprise the largest, most widespread and probably also the most uniform and typical of the groups within the babblers. Species are to be found throughout Africa, India and the Oriental region, well down into the Malaysian peninsula. Some are dull, some are colourful, but nowhere is the family better represented than in the foothills of the Himalaya. As is common among forest dwellers, the laughing thrushes are noisily vocal, and many of them are pleasantly musical, though much of the time their calls are musical chatterings designed to keep the various members of a fast-moving flock in contact with one another as they swirl through the forest like leaves swept before the wind. Like the thrushes of Europe, their diet is a mixed one of small soil and litter animals — mostly insects and worms — with fruit in season. Most birdwatchers, seeking them in this spectacular but difficult terrain, will set off after a few calls and rustles of leaves. Laughing thrushes seem always to move uphill, so progress becomes difficult and breathless through the ferns and rhododendrons, with the birds' chuckling, laughing calls apparently mocking from just beyond good viewing range!

White-crested Laughing Thrush *Garrulax leucolophus*

LILY-TROTTER

Lily-trotters, such as the African Jacana, are found in wet or swampy areas throughout the tropics. Despite their rail-like appearance as they fly awkwardly away, low over the water, legs trailing, they are more closely related to the waders than to the rails and crakes. Called lily-trotters because of their habit of stalking sedately over floating vegetation, these pigeon-sized birds achieve this feat without sinking ignominiously because their toes are extraordinarily long and widely spread, ensuring that their weight is evenly dispersed over a wide area.

The reason for their apparent sedateness of movement is partly that the feet must be carefully positioned on the floating lily leaves, and partly that the long, spidery toes can easily become entangled — even with the other foot! The plumages of the two sexes are very similar and, unlike most birds, it is the female which is slightly the larger and which takes the dominant role during the breeding season, leaving the male to cope with the bulk of the incubation on the drifting, raft-like nest.

African Jacana
Actophilornis africanus

LOON

If there were a *concours d'élégance* for birds, then a summer-plumaged Arctic Loon would be a prime contender for the title. These birds, called Black-throated Divers in Europe, have various calls — a mixture of barks, yodels and eerie wailings — but nothing as dramatic as the whooping maniacal laughter of their larger relative, the Great Northern Diver or Common Loon. The strange songs of the loons are made the more thrilling because of their remote and wild setting among mountain-ringed lakes. It is tempting to assume that the name 'loon' derives from these hysterical laughing cries, but it is more likely to originate from the Icelandic word 'lomr', meaning lame or clumsy. During their brief summer, Icelanders would be familiar with loons and well aware of their clumsiness on land as they approach their nest. These birds are supremely well adapted for life largely on or under water hunting fish. Their bodies are torpedo-shaped and streamlined, and the legs, with large feet and lobed toes, are set back near the tail for best propulsion, hence their clumsiness on land. Most divers only ever venture ashore to struggle the few feet across the bank to their nest and eggs.

Black-throated Diver *Gavia arctica*

LORIKEET

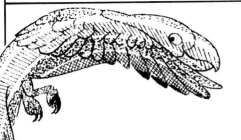

Rainbow Lorikeets *Trichoglossus haematodus*

The Rainbow Lorikeet (or Rainbow Lory) is a widespread bird in the lowland bush of Australia and the nearby islands. It feeds on pollen, nectar, seeds and fruit, with occasional insects thrown in for good measure, travelling about in flocks often dozens and sometimes hundreds of birds strong. These are noisy and ever-active flocks, shattering the peace as they frequently venture into suburban areas, screeching as they fly low over the rooftops and quickly turning and tumbling colourfully from the sky if they meet others already enjoying a good feeding location. They are hole-nesters, and, in common with other predominantly fruit-eaters, the fledging period is at least double the length of incubation, in contrast to the more usual one-to-one ratio of birds enjoying a diet richer in proteins. As their name implies, Rainbow Lorikeets sport almost all the colours of the rainbow in their plumage. To confuse the birdwatcher, there are (amazingly) at least 21 subspecies, recognisable on plumage differences (often quite striking ones), and to add insult to injury one or two of these the authorities have the audacity to describe as 'plumage variable'!

LOVEBIRD

Only 7 inches (18 cm) long, Masked Lovebirds, black-headed and with a glaring white eye-ring, come from a restricted upland plateau area in Kenya and Tanzania. To man they are among the most attractive of the parrot family, and in consequence have been introduced to other areas. Their original isolation is thought to have been due to the home grasslands, studded with acacia trees, being ringed by mountains and *Brachystegia* forest — inhospitable terrain lacking their food plants and confining the Masked Lovebird as effectively as would the sea. A loud high-pitched twittering focuses attention on their breeding colonies, where their social behaviour, particularly the habit of snuggling up to one another on a perch (so endearing to us), and their dark faces call back to mind the masked balls of olden days. Masked Lovebirds are hole-nesters, with a particular liking for the grotesque baobab or upside-down tree, but as man's influence spreads they readily build their domed nests, made of twigs and strips of bark, in cavities in buildings. So large are some of the twigs that the female has difficulty in flying straight, and watching her attempt to force an entrance (twig held crosswise in beak) to her new home is an amusing spectacle for the onlooker.

Masked Lovebirds *Agapornis personata*

LYREBIRD

Australia is an island continent of biological marvels, and paramount among the birds must stand the lyrebirds. The best known is the Superb Lyrebird. The name derives from the harp-like shape of the long outer pair of tail feathers, which are gingery brown. The central pair are just as long but slender and white, while the remaining six pairs, also long, are fine, lace-like plumes. Male lyrebirds establish large territories, perhaps a kilometre square (250 acres), and have within their territory a series of dancing arenas, about one metre wide. On these mounds they perform, hoping to attract one of the nearby females. The courtship dance has two quite extraordinary features. The first is the transformation of a dowdy, chicken-like brown bird, as the dance reaches its climax and the bird unfolds its tail, arching it forwards until the plumes reach the ground, over its back and head, in a shimmering cascade. The second is the loud and glorious song: many

birds include some element of mimicry, but the lyrebird is without equal either in accuracy or in the range of species that it imitates, blending the whole with liquid notes of its own.

Female Superb Lyrebird *Menura novaehollandiae*

MACAW

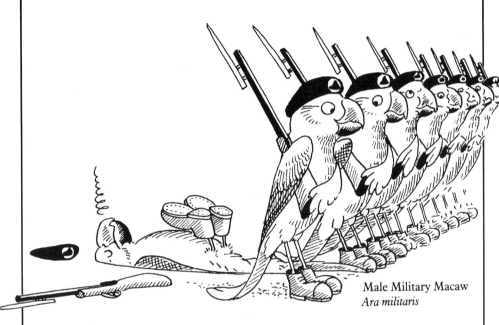

Male Military Macaw
Ara militaris

Macaws are giant members of the huge parrot family, the largest parrots in the world, in fact, averaging a metre or more in total length, much of which is accounted for by a splendidly long tail. The macaws come from jungle areas in Central and South America, but despite this remote and inaccessible origin are familiar to us all because so many are kept in zoos, where their spectacular colours and long life make them a popular feature. Brightly clad like soldiers in ceremonial uniform, Military Macaws fly in stately slow motion through the jungle, calling with a voice whose volume and raucous tone would be a credit to any parade-ground Regimental Sergeant Major. Macaws are vegetarians, so the tropical jungle, with trees of some sort always in fruit, is ideal habitat. But fruit is high in volume and low in nutritional value, so the young may be in the nest — a huge hole in a tree — for upwards of three months before they are able to leave and fly free.

MANDARIN

Male Mandarin Duck *Aix galericulata*

Although there is a temptation to assume some culinary link between the conspicuous orange feathers of the male Mandarin and duck *à l'orange*, the gaudy plumage of the Mandarin Duck, owing to its similarity to the richly-coloured silken robes of the Chinese dignitaries, is more likely to have given the bird its name. Mandarins are plump and neatly small as ducks go and, rather than waddling as do many of their relatives, have a short stride so that they walk with small steps, again reminiscent of the mandarins from the Orient. Their origins do lie in eastern Asia, from where the first pairs were imported to beautify ornamental waterfowl collections in Europe. Since the first introductions in about 1750, pairs have escaped and small free-breeding feral populations exist in several areas.

MARTIN

House Martins
Delichon urbica

House Martins are best known as summer visitors, building nests of mud beneath the eaves of houses. Often the origins of the nest can be traced by the colour of the little pellets of mud (the 'bricks') of which it is made, and the builders can be followed and observed, standing delightfully and delicately on tip-toe on the muddy margins of a puddle, attempting to avoid getting their feathered feet muddy. House Martins are prime examples of birds that have benefited from man and his works: in days long past they must have been cliff-nesters of restricted distribution, but they have taken readily to houses, and especially to the clinically constructed eaves of modern private housing estates, which otherwise would be pretty sterile for bird life. Though numerous in Europe in summer, in winter the House Martin is an enigma, for, despite many ringing studies and much observation, astonishingly little is known of their whereabouts or habits in Africa.

MEADOWLARK

There are two species of meadowlark in North America, Eastern and Western. They are elegant birds, supremely well camouflaged from above, their backs a streaked mixture of browns and fawns ideal for concealment among the dry grasses of the meadows, plains and prairies that are their home. Usually they are flushed at startlingly close range, rising on fluttering wings and conspicuous only because of the white flashes in the tail. Surprisingly, when walking about, the tail is continuously flicked open, like a fan, exposing these glaring patches to the full and spoiling the camouflage effect. Should the meadowlark perch on a post or bush, then the beauty of the rich yellow throat and breast, separated by a broad black 'V' collar, can be appreciated to the full. Pure geography is enough to separate the two species at east and west extremes, but in areas of uncertainty song is the deciding factor. The Western Meadowlark has a bubbling flute-like melody, whereas the Eastern can manage only a simple double-barrelled whistle. It is difficult to see how such differences could arise in species so similar: it would be a brave man who suggested that the cause was a lack of competitive musical talent in the eastern states! The Red-breasted Meadowlark is an unrelated species from South America.

Red-breasted Meadowlark
Leistes militaris

MERLIN

Fast and agile, the Merlin is a compact falcon, specialising as a hunter of Meadow Pipits and other small birds in its moorland habitat. Unlike its larger and more powerful relative, the Peregrine, which hunts by 'stooping' from high in the sky onto unsuspecting quarry, the Merlin is a low-level striker, overhauling its prey from behind and sometimes below, relying heavily on surprise. Merlins have no connection with the Welsh magician of the same name (and similar habitat), except that they have the ability to disappear with incredible swiftness into an apparently featureless landscape — much to the annoyance of birdwatchers. The name is likely to have been derived from the French *merle*, meaning Blackbird — the maximum-sized prey a Merlin was likely to tackle when set to hunt by a falconer.

Merlin *Falco columbarius*

MOCKINGBIRD

Mockingbirds are insect- and fruit-eaters, and excellent songbirds, perhaps among the best songsters in the New World. They are the more attractive to man because they frequent bushy open areas where town and country merge, penetrating readily into garden habitats. The song of mockingbirds is varied and long-running, and contains a great deal of mimicry — often repeated — of snatches of other birds' songs: hence 'mocking'. On the Galapagos Islands, off equatorial South America, the mockingbirds are fascinating. In beak length and plumage they vary intriguingly from island to island, though often these are only 30 or 40 miles (50-65 km) apart: an indication of the mockingbird's terrestrial habits and rather feeble flight capability. On Hood, one of the most arid islands in the group, evolution has endowed the mockingbirds with extra-long beaks, which they use to pierce the emerging feather quill-sheaths of incubating boobies, sucking at the blood as a source of moisture. They have a devil-may-care attitude to man, and show no hesitation in alighting on the rim of a cup and drinking from it as it is put to your lips, especially if the contents are fruit juice.

Galapagos Mockingbird *Nesomimus trifasciatus*

MOLLYMAWK

The name 'mollymawk' over the centuries has been applied by mariners to a variety of seabirds, but perhaps most often to the Black-browed Albatross. The name is derived from the Dutch — a seafaring nation — *mal* for foolish, *mok* for gull. Obviously, in the days of sailing ships, seabirds were welcome fresh-meat additions to enliven a monotonous diet, and those that sat fearless on their nests — as albatrosses would — while the sailors slaughtered as many as they needed naturally became well known. Though habitually birds of the wide open southern oceans, Black-browed Albatrosses do occasionally stray into the Northern Hemisphere. Here they tend to link up with Gannets — perhaps the nearest equivalent in size, shape and habits that they are likely to find — on a more or less permanent basis. They will even set up lonely territory in the midst of a gannetry, and records stretching back for some centuries indicate that this is not a modern phenomenon, as the albatross in times past was called 'King of the Gannets' when present in their colonies.

Black-browed Albatross *Diomedea melanophris*

MOUSEBIRD

The mousebirds (or colies), of which the Speckled Mousebird is the best known, are a drab and untidy-feathered family of African birds, rather resembling middle-sized mouse-coloured long-tailed parakeets. They are widespread in open bush areas, and have taken readily to small farms and gardens, where their flexibility in dietary requirements and parrot-like beak enable them to come to grips so quickly with all types of food from seeds and fruit to greenstuffs that they are often regarded as pests. So clumsy are they in flight that for them the statement 'flew into the bush' is literally true. They move about in small gangs, and should a bush intervene in their erratic flight path they may crash into it, and hang at all angles from whatever leaves, twigs or thorns their feet may have grasped. After a few moments' respite, some will right themselves and begin to feed, but others will remain as they arrived, sometimes extending their wings to sunbathe, when a contentedly glazed expression replaces the look of consternation on their faces.

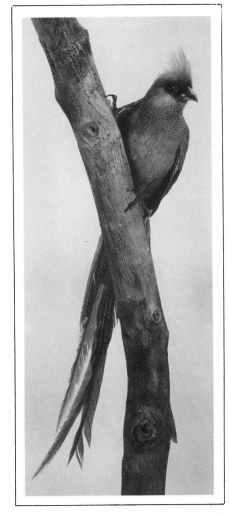

Speckled Mousebird *Colius striatus*

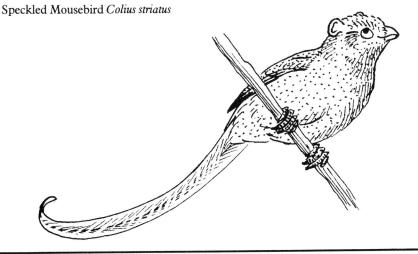

MYNA-BIRD

In captivity the glossy coal-black myna, with its strangely fleshy golden-orange beak and yellow head patches, is often regarded as one of the best talking birds, so good a mimic is it. By nature, many of the myna's calls are throaty or chuckly, which gives a particularly lewd slant to some of their mimicry, particularly in pubs. Sadly, the pressures of hunting for caging are beginning to affect numbers of Hill Mynas (for this is the species usually caged), which in some areas is a threatened species, not least because so many of those caught in the forests with primitive techniques such as bird-lime fail to survive the rigours of the journey to captivity in the so-called civilised areas of the world. In captivity, too, their plumage loses much of the deep purple sheen to be seen in their native habitat, the densely wooded country of the foothills of the Himalaya and the mountain ridges of southeast Asia.

Hill Myna *Gracula religiosa*

NIGHTINGALE

For a bird so seldom seen, the Nightingale is astonishingly well known to modern man, perhaps because its song is one of the most fabulous anywhere in the world. In the past, too, it was much appreciated, the Ancient Greeks being fulsome in their praise. There are several traditional fables accounting for its nocturnal song. One suggests that it sings with its breast against a thorn to keep awake (the legendary 'thorn-bird'). Oddly, the reason for this wakefulness is supposed to be that the Nightingale, originally endowed with only one eye, stole the single eye that legend also held the slow-worm to possess. The slow-worm is now on a vengeful hunt for the Nightingale, which sings all night to keep awake. Hence the fascinating collective noun for a group of Nightingales — a 'watch', as in night watchman or the watch kept by sentries or sailors.

Nightingale *Luscinia megarhynchos*

NIGHTJAR

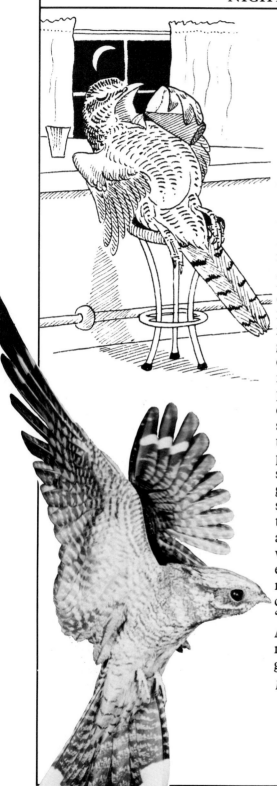

Nightjars are beautifully camouflaged summer visitors to open woodlands, usually on drier sandy soils, particularly favouring the rapidly diminishing heathland areas and sections of recently-felled woodland. Their camouflage is as effective when they are sitting out the daylight hours perched horizontally on a branch as it is when they are incubating the eggs (also well camouflaged) in a ground nest that may be situated against backgrounds as diverse as oak-leaf litter or the ashes of a brush-clearance fire. With their huge mouths, Nightjars have been described as 'aerial plankton feeders' — catching as many insects by sheer size of gape as by good nocturnal vision. 'Night' is an obvious reference to the bird's nocturnal habits, while the 'jar' probably relates to an eighteenth century word for a 'quivering sound', an admirable description of the Nightjar's long-drawn-out purring song. This noise, and the sudden appearance of Nightjars giving wing claps in display sufficiently frightened countrymen to give rise to a rich folklore. Puck, a mischievous fairy, was associated with the Nightjar, but moreover it earned an (absolutely undeserved) reputation as a stealer of milk from cows or goats in the field — hence 'Goatsucker'. This goes back to Aristotle — and indeed the generic name *Caprimulgus* means, literally, goat-sucker.

Male Nightjar *Caprimulgus europaeus*

NODDY

Though similar in anatomy to the other members of the tern family, the noddies form a distinct little group of four species, of which the Brown Noddy is the most numerous and widespread. For a start, they are all drably and darkly plumaged, three being dull brown, the other grey. All inhabit tropical or subtropical seas, breeding on the numerous oceanic islands in this circumglobal belt and rarely travelling more than a few miles out to sea from their colonies. But strangest of all are their nesting habits. Terns, almost exclusively, are ground-nesters, making a perfunctory scrape in the sand and relying on cryptic egg and chick coloration for the safety of the nest contents. The noddies, although occasionally nesting on ledges within caves, usually build nests of a few twigs and seaweed up in the trees, and a colony in full swing gives the birdwatcher's preconceptions something of a shock. Here the male performs the vigorous head-nodding display that provides a name for the group. One other unusual feature remains: noddies catch most of their fish almost at second hand. Their usual prey is small fry, caught as they leap and skip out of the surface when chased from below by bigger predatory fish, so only rarely does the noddy dive and get wet.

Brown Noddy *Anous stolidus*

NUN

Presumably because of the resemblance of its black hood to the ecclesiastical head-dress, the Black-headed Mannikin or Munia is often called the Nun by cagebird fanciers. Indian in origin, this seed-eater is a popular bird in captivity because of its elegant plumage. Despite a very bulky beak for its size, it favours small seeds, especially of grasses, and especially those just beginning to germinate. Nuns are flocking birds, feeding gregariously and building their oversized bulky twig nests in loose colonies. The nests are domed as an additional protection against predators to the spiny bushes in which they are usually built. Though so widely kept in captivity, sadly most Nuns are taken from the wild as this is reputedly a difficult species to breed. This may be because they are all too easily trapped and are readily available in bird markets, or it could be due to the problems that aviculturists encounter in getting the sex of their birds right, as the two have identical plumage!

White-headed Nun *Lonchura maja*

NUTCRACKER

The nutcrackers are Jay-sized birds, related to the crows, dwelling in the mountainous conifer forests of Europe and North America. The European birds are generally brown in plumage, flecked with white, but Clark's Nutcracker, a bird of the Rocky Mountains of North America, is a soft grey with strikingly black-and-white wings and tail. It was discovered by the Lewis and Clark expedition early in the nineteenth century and named after Clark himself. It is an omnivorous bird, like all crows, eating insects and their larvae, the eggs and young of other birds, but specialising in nuts and seeds. They are experts at hacking open fir cones to extract the seeds, holding the cone in one foot while they hammer away at it with their beak. As befits birds that must survive a harsh winter climate, the nutcrackers are great food-storers, taking nuts away and burying them as a cache for future emergencies.

Though not infallible, they are much better than squirrels at refinding their stores, even when the ground is covered in a blanket of snow.

Clark's Nutcracker *Nucifraga columbiana*

NUTHATCH

nestbox lid may be cemented firmly on in this way, and oversized entrance holes are plastered around inside to reduce them to the correct diameter. They may use astonishing quantities of mud: in southern England, some nests in cavities in haystacks were reported to have had several pounds of mud to reinforce the fragile entrance!

Nuthatch *Sitta europaea*

Although they do take insects and their larvae, particularly in summer, Nuthatches specialise in nuts. Larger and harder ones, such as chestnuts, beechmast and hazel nuts, are carried in the beak to a suitable (and regularly used) crevice in the bark or to masonry of a nearby building and hammered open with the beak. This characteristic hammering is one of the best ways of locating them in winter. In many areas, Nuthatches are regular and popular visitors to the garden bird table, where they eat most offerings including fat. Often they will take away tougher items such as peanuts (or even hard toast) and subject them to the anvil treatment on their favourite trunk. They nest in holes or cavities, often far larger than necessary, flooring the nest chamber, very characteristically, with flakes of dry bark. Another characteristic of all Nuthatch nests, even those in nestboxes, is the liberal plastering of mud that seems part of the essential pattern of Nuthatch life. A

OWL

The glaring orange eyes of the Long-eared Owl are given additional ferocity by the face markings and the erect feather tufts — the 'long ears' — on the top of the head. These are not in fact ears, but simply feather adornments, functioning in display. Nonetheless, Long-eared Owls have phenomenal hearing powers. They are largely nocturnal hunters of small rodents and birds, with the huge eyes and exceptional poor-light vision common to most of their kind. The disc of feathers surrounding the eye is hard to the touch, unlike the velvety feel of the rest of the owl's plumage, which effectively acts as a silencer as it flies. This disc serves like a parabolic reflector, as used in radar or by sound-recordists, and collects sounds, focusing them on the huge ears concealed behind the head feathers. These ears are asymmetrically positioned on the skull, for better range-finding, and enable Long-eared Owls to strike and kill accurately in total darkness. With so much of the skull occupied with specialised and huge eyes and ears, there is little room left for the owl's legendary wisdom!

Long-eared Owl *Asio otus*

OYSTERCATCHER

Among the most conspicuous of shorebirds with their pied plumage, brightly coloured appendages and continuous piercing calls, few Oystercatchers today have the chance to feed on oysters. There seems to be something of a division of thought among Oystercatchers as to what is the most appropriate prey. Some now live away from the rocks, even in damp meadows well inland, and these use their beaks to probe for worms — and train their chicks to do the same. Those still favouring the rocks tend to specialise in limpets, which are approached by stealth and dislodged with a hammer-like tap of the beak, or on bivalve molluscs such as cockles or mussels, which are obtained by probing in the sand or picked off the rocks. Limpet 'specialists' tend to have shorter, stouter beaks than the others, which use the fine chisel end of the beak to open the bivalves, then snip out the contents with a scissor-like action. Again, the youngsters are 'trained', learning by experience, to follow in their parents' expertise. Adaptable birds when it comes to nesting sites, too, Oystercatchers choose fields, sandy beaches, estuary islands (and even post-tops in decaying jetties) and, most recently, gravel-covered flat roofs of large factories!

Oystercatcher *Haematopus ostralegus*

PADDY BIRD

The Paddy Bird, or Indian Pond Heron, is the Asian equivalent, both ecologically and in its almost identical plumage coloration, of the Squacco Heron of Africa. It is a bird of roadside ditches, pools and rice paddies, as its name implies, and most remarkable for appearing almost wholly white in flight, so round and moth-like are the white wings. On landing, these wings are folded rapidly out of sight beneath the long, loose, feathers of the back, and so well does the buff streaky body match its surrounds that the bird seems to vanish instantly. Equally startling is when it takes suddenly to flight. Paddy Birds show no signs outwardly of being bad-tempered, nor of any Irish descent!

Indian Pond Heron *Ardeola grayii*

PAINTED SNIPE

Painted Snipe are unusual and distinctive wading birds, in many ways combining the plumages and behaviours of two other groups — the true snipes and the rails — that live in similar habitat, the muddy reed-fringed edges of tropical wetlands. In needing to be almost trodden on before they will flush, they resemble the snipes, but it is particularly in flight, when the short rounded wings and legs held trailing can be seen, that the Painted Snipe most resembles a rail. As with one or two other wader species, Painted Snipe are unusual in that it is the female who is the more brightly coloured and who takes the lead in display. After mating, she is unable to avoid the chore of actually laying the eggs, but once this is accomplished she leaves the incubation and hatching of the young entirely to her much dowdier spouse.

Painted Snipe *Rostratula benghalensis*

PARAKEET

Rose-ringed (or Ring-necked) Parakeets are among the more numerous of the parrot family in Africa and India. In consequence, many are caught as cagebirds and exported elsewhere — a traffic that has little appeal in a conservation-conscious society and a wind sown which, when reaped, may turn out to be the proverbial whirlwind. Even as far from their native land as Britain, Rose-ringed Parakeets which have been released (or escaped) from captivity into the wild are establishing themselves effectively. In their natural surroundings, Rose-ringed Parakeets eat all manner of seeds and fruits, including a wide range of crops, which brings them into conflict with man. Much the same is true in Britain, although it seems to be apples rather than cereals that are the prime target — but it is early days yet! Originally it was thought that the severe northern winter would quickly eliminate these feral birds, but, aided by the British habit of putting out food for the birds in bad weather, they are surviving to the extent of several hundred breeding pairs — a success story, but with sinister overtones.

Rose-ringed Parakeets *Psittacula krameri*

PELICAN

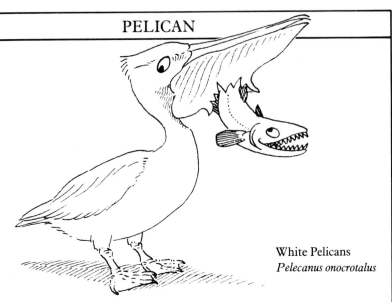

White Pelicans
Pelecanus onocrotalus

Along with the owl, the pelican shares the distinction of being familiar to most of us from our earliest childhood upwards, an extraordinary, slightly grotesque bird typified for example by Edward Lear: 'whose beak can hold more than his belly can'. There are other, less precise, references to pelicans and their supposed habits stretching back to pre-Christian times. The real pelicans though are birds of the warmer parts of the world, favouring lakes and rivers rich in fish. Despite their size, they are capable of soaring effortlessly on their long broad wings, and derive much assistance from the thermals or hot-air upcurrents rising from the sun-baked landscapes. Although, despite their size and apparent ungainliness, some pelicans dive like gannets for food, the White Pelican is a more sedate fisherman. Often a group of half a dozen or more will co-operate, swimming along in a horseshoe formation. When they have a school of fish trapped, they close up into a circle, and as one dip their huge pouched beaks under the water, lift them, again in slow-motion unison, and swallow the catch before swimming on in stately fashion to fresh hunting grounds.

PENGUIN

Macaroni Penguins nest in the most enormous colonies — called, strangely enough, rookeries — on islands around the Antarctic continent. These colonies are usually sited on flat or gently sloping rock or earth, and may be hundreds of thousands of birds strong. Not unnaturally, not only is the sight and sound as spectacular as any in the bird world, but the smell — or better, stench — arising from their guano, decaying fish and rotting vegetation is quite indescribable. Each yellow-crested pair of Macaronis forms a crude nest of pebbles, in which the female lays two eggs. The first is small, and normally discarded after a few days, while the second is incubated, all going well, to maturity. Though clumsy on land as they waddle, often with comical apparent urgency, to and from the nest — a journey sometimes of several miles — and as they flop gracelessly down onto their egg, in the sea all penguins are in their element. Streamlined, insulated by a thick blubber layer and propelled by powerful flippers, they often exceed 30 mph (50 kph) below water and frequently 'porpoise' in and out of the waves, arriving back ashore in a hilarious somersaulting tangle of feet and flippers among the breakers.

Macaroni Penguins *Eudyptes chrysolophus*

PINTAIL

There can be no doubt as to the origin of the name of the Pintail, as the drake has one of the longest tails in the duck family, and even the sombre cinnamon-brown female has a relatively elongated tail. Over Europe as a whole, the slim and extremely elegantly plumaged Pintail is one of the most numerous and the most widespread of the dabbling ducks. Most dabbling ducks, beside squidging their beaks through the mud at the water's edge, also tend to up-end in the shallows, reaching down to feed on aquatic plants below the water. Long and slim in the neck as well as in the tail, the Pintail — which looks when feeding in this way very like a fisherman's giant float — can outreach all of its relatives.

Male Pintail *Anas acuta*

PLOVER

Conspicuous in its grey, black and white plumage, the Blacksmith Plover is widespread and often quite common in east and central Africa. Though primarily a bird of the margins of lakes and rivers, it often nests some distance from water and has adapted well to agricultural land. The name is derived from the call, a solid metallic tapping resembling the clang of hammer on anvil in the old forges of blacksmiths. For many birds, especially ground-nesters like the plovers, the baking heat of the sun in tropical Africa presents problems. Birds, being feathered, cannot sweat and thus have problems losing enough heat by panting; to assist, the incubating bird sits tail to wind, with its feathers raised to deflect cooling airs down onto its skin. For the Blacksmith Plover and others, 'incubation' may often be more of a 'sunshade' task during the day, the bird standing over its eggs and just shading them from the direct sunlight. Even standing can become a problem, and Blacksmith Plovers counter this by squatting on their 'haunches' (actually their ankles) to lift their toes clear of the burning hot sand.

Blacksmith Plover *Anitibyx armatus*

PUFFIN

Some Puffins nest in holes in cliff faces, and large colonies may occur in screes at the head and foot of the cliffs, but most nest in grassland burrows. These they may dig for themselves, or commandeer from rabbits or Manx Shearwaters. The slopes are usually covered in pink thrift, white sea-campion and often bluebells, all flourishing on the guano deposits and forming an attractive backdrop to the communal fly-pasts and social gatherings that are so much a part of a Puffin's life. The huge parrot-like beak, so colourful with red, blue and yellow bands in summer, grows only for display in the breeding season: at the end of the season it is replaced by a much smaller, strictly functional, grey winter beak. Once the young have hatched in June, adults visit their burrows with sand eels or fish fry as food. Their carrying capacity is phenomenal: often there may be a dozen or so fish held crosswise, but up to 60 small fry in one beakful is on record. Puffins have the flight capability of the average brick and find difficulty landing in strong cross-winds, a fact exploited by Herring Gulls, which harass the Puffin into a crash-landing, when the fish it drops are gobbled up by the gull.

Puffin *Fratercula arctica*

QUAKER

Also known as Monk Parakeets, Quaker Parakeets are commonly kept in captivity. Long-tailed, stoutly built birds, they originate in the open forests of the southern part of South America. Despite the size of their broad, strong beaks, they can handle a wide variety of food from the tiny seeds of grass and millet, through rice and potatoes to sizeable fruit. In recent decades, extensive planting of trees such as eucalyptus has helped the Quaker Parakeet extend its range well to the north in South America, and other populations originating from escaped cagebirds now flourish in Puerto Rico and, astonishingly, in the northeastern USA. They build large, twiggy domed nests, like Magpies choosing the topmost and safest twigs of the trees, usually nesting in colonies. Unlike the Quakers, and particularly unlike most orders of monks, they are extremely gregarious and noisy, often squabbling furiously over a nest site or food titbit. Just as un-Christian is their habit of descending in droves onto man's crops, ranging from maize to citrus fruits.

Quaker Parakeet *Myiopsitta monachus*

RAZORBILL

Any bird-ringer (or bird-bander, for that matter) who has inadvertently handled his first adult Razorbill without taking the precaution of wearing gloves can give a ready definition of the origin of the name. The edges of the beak are remarkably sharp, and that heavy-looking skull is bulky with the muscles that power the bite. When it comes to feeding, though, these North Atlantic seabirds can be delicacy itself. Maritime for much of the year, they come ashore only to breed, choosing protected sites in nooks and crannies on the cliff face or beneath boulders at its top or foot. The single chick is fed with small fish, supplied by the beakful, and each beakful may contain a dozen or more small fry. These are all held crosswise in the beak, despite its power and sharpness, and, although usually battered, are rarely cut through. Razorbills are smart rather than ornate birds, their black heads enlivened by only a white

horizontal line from eye to beak and a white vertical line *on* the beak, near the tip. It has been suggested that a possible purpose of these is as sights for underwater hunting: the horizontal is the sighting line, and when a fish nears the vertical line a snap of the beak will secure it!

Razorbills *Alca torda*

ROBIN

The Robin was voted British national bird some years ago — deservedly so, as its popularity extends from the garden to the cards on the mantelpiece at Christmas. At the turn of the century, the old English 'Robin Redbreast' was abbreviated to just 'Redbreast' as the commonly used name, but this has now been superseded by the more popular contraction 'Robin'. Robins have always been at the centre of superstition, and to kill one was regarded almost as sacrilege, a feeling perpetuated even now in the nursery rhyme 'Who killed Cock Robin?'. These beliefs, and the Robin's popularity as a Christmas-card feature, stem from the legend that the Robin attempted to draw the nails (or, alternatively, remove the crown of thorns) at the Crucifixion, receiving for its efforts the drop of Christ's blood that forms its red bib. Beneath a friendly exterior, the Robin is a tough little customer. The two sexes are alike in plumage, and even Robins seem able to tell who's who only on the basis of behaviour. In winter, male and female set up separate feeding territories and she will hold her own just as well as he, hurling abuse at any invader and chasing it off.

Robin *Erithacus rubecula*

ROLLER

As their flashing, jewel-like plumages would suggest, most of the roller family are birds of the tropics, though some, including the European Roller, penetrate into warm temperate regions in both hemispheres. Mostly pigeon-sized, rollers concentrate on ground prey ranging from lizards and other small reptiles to insects — mostly large ones. A typical hunting technique is to choose a perch overhanging an open grassy area or track, and to drop in a flash of colour onto prey passing below. Often the prey is awkward and leggy — stick insects and grasshoppers for example — but the capacious gape of the roller can eventually get round the most obstinate of meals. In the breeding season, rollers have a wonderful tumbling display flight which includes mid-air rolls that display their azure-blue wings to best effect.

Roller *Coracias garrulus*

ROSEFINCH

Though his female must surely be a front-runner in any competition to select the least memorable bird in the world, the male Scarlet Rosefinch (or Grosbeak) is one of the most colourful of the finches, outshone only slightly by the related Pallas's Rosefinch of Siberia and the Great Rosefinch of the Caucasus. It is a bird of northern Eurasia, nesting in the birch and aspen woods beside the rivers running through the taiga and extending up into subalpine habitats in the foothills of the Himalaya. Often it is to be found associated with human cultivation — perhaps (and paradoxically) because the broken ground that results from agriculture (in areas where chemical and mechanical aids to farming are too expensive to be commonplace) gives rise to a richer flora and thus a greater abundance of weed seeds. In many areas, though hardly popular as a result, the Scarlet Rosefinch (whose beak structure resembles that of the Bullfinch, and it has similar feeding habits) can survive the winter by eating buds from trees, finding — not unnaturally — that the bigger buds of cultivated varieties such as apples and pears offer a better meal than wild alternatives.

Female Scarlet Rosefinch
Carpodacus erythrinus

RUBYTHROAT

Once called the Rubythroat Nightingale because of its white-edged ruby-red throat patch and Nightingale-like song, the Siberian Rubythroat is a Robin-sized bird characteristic of the taiga areas of Russia and China. It is a bird of dark, swampy conifer forests, normally with a tangle of fallen trees and a consequentially well-developed and almost impenetrable undergrowth. Its strong legs, a little oversized for its body, indicate a terrestrial life style that befits its habitat and its major foods, which are thought to be the various insects, soil arthropods and molluscs, and worms that live in the moss and leaf litter on the ground. The Siberian Rubythroat has a beautiful song, rarely heard by appreciative birdwatching ears because of the remoteness of its breeding grounds. It is a skulking bird, so in addition only rarely is it well seen, even when it migrates south for the winter, as it tends to move towards southeast Asia where birdwatchers are few.

Siberian Rubythroat *Luscinia calliope*

RUFF

Ruff *Philomachus pugnax*

A male Ruff (the female is normally called a Reeve) in full breeding plumage would be enough to excite even the most exotically dressed Elizabethan courtier, after whose pleated neckwear the Ruff is named. Unlike the neatly folded ruffs of the Elizabethans, the avian Ruff has a band of feathers around the base of the skull of a wide variety of colours. Some, as in their human counterparts, are white; others brown, gold, chestnut or black, or a mixture. This plumage is for display only: on their wintering grounds in Africa the males are as nondescript as the smaller females are year-round, obviously for better camouflage protection. Display takes place at a communal 'dancing ground' or lek, where rivals posture, ruffs extended, and indulge in mock (and sometimes actual) battles to attract the females' attention. Strangely, white-ruffed males, though conspicuous, are rarely if ever successful suitors.

SADDLEBACK

The Saddleback is one of a group of three New Zealand forest birds, the wattlebirds. Best known, but now extinct, is the Huia. So distinct were the beak shapes of male and female that for many years museum experts (lacking any knowledge of the Huia in the wild) considered them to be of separate species. Saddlebacks are smaller than Huias, about 10 inches (25 cm) long, with generally glossy-black plumage and a strikingly contrasting chestnut back. The sexes are alike. In flight Saddlebacks are feeble, preferring to flit from branch to branch through the forest, the main propulsion coming from their powerful legs and feet. The sharp, woodpecker-like beak is used to catch insects and to prise them from nooks and crannies, but they also relish fruit and sometimes delicately (for their size) sip nectar from large flowers. Although scarce and confined to a few offshore islands, Saddlebacks seem literally to be enjoying the protection of conservationists, and appear to be flourishing — rather than following the Huia into oblivion, save in the museums of the world.

Saddleback *Creadion carunculatus*

SANDPIPER

Common Sandpiper
Actitis hypoleucos

The eggs, and the downy newly hatched young, of the Common Sandpiper are among the best camouflaged of any wader — and the waders as a family excel in nest concealment. The nest is usually in dead vegetation or among the pebbles on the banks of a fast-moving clear stream, and the female sandpiper will sit tight, her speckled plumage giving her protection, until the last minute as a potential predator approaches. If that 'predator' is a birdwatcher, having a Common Sandpiper shoot up and whirr away from almost under your feet is a heart-stopping moment. Earlier in the breeding season, their behaviour could not have been more in contrast. Anxious to secure his territory and attract a mate, the male will have patrolled his stream incessantly, flying low on shallow-beating wings over the *sand*banks, uttering his *piping* trill almost without ceasing — hence the name.

SECRETARY BIRD

At first sight, striding across the grassy African plains, the Secretary Bird is difficult to place in a family. Seen overhead in flight, its silhouette is so heron- or stork-like as to be even more confusing. But Secretary Birds are considered to be aberrant relatives of the birds of prey, and a glance at their eagle-like beak and eyes confirms this. Their long legs are an adaptation to unusual prey: they hunt snakes (and other reptiles and large insects) on the plains, and the legs give them not only speed covering the ground but a high vantage point. When a snake is spotted, the legs are again useful, for the initial attack is with fiercely kicking feet, the body of the Secretary Bird well out of range of venomous fangs. Their name arises from the untidy crest of feathers atop the head, looking like the quill pens tucked behind the ears of a secretary a century or more ago.

Secretary Bird *Sagittarius serpentarius*

SEEDEATER

Were there to be a title 'least conspicuous bird in the world', or perhaps 'least glamorous', then the Streaky Seedeater would be a front-runner from the off. Seedeaters are smallish African finches with a relatively heavy beak, giving the overall impression of off-colour female sparrows. Above they are drab brown with darker markings, below dirty buff with drab brown streaks — hardly contenders for the *concours d'élégance*. They are, though, placed in the genus *Serinus*, in company with the various wild members of the canary tribe with their golden plumages — and to hear a Streaky Seedeater in song is to understand why. Rather on the lines of Hans Andersen's swan, this dowdy bird possesses the beautiful extended bubbling song of its relatives.

Streaky Seedeater *Serinus striolatus*

SHEARWATER

Manx Shearwaters are slender-winged seabirds, the size of a small gull, which nest underground on remote and usually uninhabited islands. As their name suggests, at sea they fly low, often touching the waves with one wingtip as they bank and turn. Their powers of navigation are formidable: a bird removed from her egg in a Welsh burrow, taken to Boston, USA, by plane and liberated was home before the letter announcing her release! By day, walking the cliff top, there would be no reason to suspect a colony in the 'rabbit burrows' all around. At night, the adults' eerie caterwauling as they fly overhead with a rush of air like a wartime shell gives the game away. The tremendous cacophony of large colonies is described in Icelandic *Sagas*, and in Norway in the past the mysterious and terrifying noises were attributed to trolls.

Manx Shearwater *Puffinus puffinus*

SHEATHBILL

The sheathbills are extraordinary birds, the centre of some dispute among bird taxonomists and very confusing to the birdwatcher when first encountered. They are thought to provide a 'missing-link' connection between the otherwise seemingly difficult-to-relate gulls and waders, being able to swim buoyantly, having slightly webbed feet but a well-developed hind toe. The field problem is that they most clearly resemble a rather eccentric white pigeon as they strut lugubriously about on sub-Antarctic beaches. Though snow-white in plumage, this is far from true of the character of their behaviour. It would be fairest to describe the Snowy Sheathbill primarily as a scavenger, obtaining food value from the most appalling offal, refuse and even seal excreta. More sinister, they wander about (and it is difficult to avoid the adverb 'slyly') among breeding penguins, attacking and sucking out the contents of any unattended egg and frequently making off with, then killing and eating, newly hatched penguin chicks.

Snowy Sheathbill
Chionis alba

SHOEBILL

There are three strange families allied to the storks and herons, each distinctive in its own character but none really closely fitting in with the rest of the group. They are the Boatbill (close to the herons), the Hammerhead (some similarities with both herons and storks) and the Shoebill, which is perhaps nearest to the storks. The Shoebill, sometimes called the Whalehead Stork, is a strange, rather grotesque-looking dweller in the papyrus swamps of eastern Africa. It is shy, despite its size (standing about 4 feet — 1.2 metres — tall), and solitary, usually taking flight at the sight of man. Its name is derived from its deep, broad beak, which seems so heavy that it often rests on its throat, giving the Shoebill a distinctly doleful expression. This huge beak is adapted for feeding in shallow muddy water, and among favourite foods are said to be items as diverse as lungfish and baby crocodiles and turtles, together with the more predictable frogs. Although anatomically closer to the storks, the Shoebill shares with the herons the habit of flying with the neck folded back, and the possession of powder down (specialised feather patches giving a powder for use in feather care) and a comb-like central claw, again for feather care. Cosmetically far better equipped than its appearance would seem to justify!

Shoebill *Balaeniceps rex*

SHOEMAKER

One of the commonest of the petrels of southern oceans is the White-chinned, also called 'Shoemaker' and 'Cape Chicken'. The last name is to distinguish it from the Pintado Petrel or 'Cape Pigeon', which is pigeon-sized and, like the Shoemaker, an habitual follower in the wake of boats. The Shoemaker is almost chicken-sized, a bulky heavyweight for a petrel, and will persist in following ships for many miles. Occasional food benefits are to be had from scraps thrown overboard, but it is thought that the turmoil of water in the wake reveals planktonic animals that would otherwise remain unseen.

Shoemaker *Procellaria aequinoctialis*

Shoemakers breed on remote southern islands such as South Georgia, nesting in burrows but strangely building a tall, bulky, gull-like nest down the tunnel. Although normally sooty-plumaged, with the white chin of their name, Shoemakers retain the ability to confuse even seabird experts by occasionally appearing in a plumage lacking the white throat patch.

SHOVELER

So conspicuous a bird is it, and so frequently does it indulge in its feeding habit of squatting on its belly in the soft ooze at the water's edge shovelling up beakfuls of mud, that it is not difficult to see how the Shoveler got its name. The spelling, though, does take some explaining. Shovelers have relatively huge beaks for dabbling ducks, spoon-shaped and equipped with a complex array of comb-like channels, grooves and bristles. The mouthful of ooze that the bird picks up is squeezed out over this sieving device, which effectively captures reed seeds and small animals while the mud is returned to the water. Not only does the beak look disproportionately large, it also makes the Shoveler float head-heavy on the water, and seems to make the wing beats desperate in an attempt to keep the Shoveler airborne.

Female Shoveler *Anas clypeata*

SKIMMER

The African Skimmer is one of three separate but similar species of skimmer that occur in tropical areas of the Americas, Africa and Asia. All are characterised by the grotesquely elongated, narrow and deep-keeled lower mandible of their beaks, which is often twice the length of the tern-like upper mandible. Not for them are the open seas, as their hunting technique would be interfered with by wavelets and destroyed by rough water. These are birds of the smooth waters of estuary or lake, where they hawk low over the surface in steady, straight flight. The long lower mandible touches and 'creases' the water surface in a straight line — hence the name skimmer, and the fact that their wing beats do not go below the horizontal. As soon as the beak touches food, it is quickly snapped up, and somehow the skimmers manage to avoid breaking their necks in collisions with heavy floating objects. Much of their hunting is done in the evening or by moonlight: the line of broken water left by the beak stimulates phosphorescent glowing in some of the plankton, and other small creatures rise to eat this plankton. Turning often 'on a sixpence' to snap them up, it is these small fish and shrimps on which the skimmers feed.

African Skimmers *Rynchops flavirostris*

SKUA

As its name implies, the Arctic Skua breeds in a circumpolar belt of islands and tundra. Skuas are related to the gulls and terns, and show a combination of the flight attributes of both, coupling the power of the gulls with the speed and agility of the slender-winged terns. This suits their life style well, as Arctic Skuas genuinely live up to their specific epithet *parasiticus*. They use their flight skill and agility to harass remorselessly homecoming gulls, Kittiwakes especially, and various terns, chasing them about the sky, sometimes in small teams, until the unfortunate parent is forced to disgorge the fish meal it is taking home to its chicks. So swift are the skuas that the dropped food is normally caught before it reaches the waves. On their own breeding grounds, Arctic Skuas are determined in defence of their nests, and will dive-bomb even human intruders with devastating effect, drawing blood from many an unprotected scalp!

Arctic Skua
Stercorarius parasiticus

SKYLARK

'Hark! Hark! The Lark at Heaven's gate sings' wrote Shakespeare in Cymbeline. Surely no noise is more typical, more redolent, of the open countryside in summer than the song of the Skylark. Often this is produced at several hundred feet (not quite Heaven's gate perhaps) with the songster very difficult to locate against a bright blue sky. If you can find the singer, and study him through binoculars, watch how precisely he maintains his station, hovering on quickly beating wings, tail fanned, bobbing about as if attached to the sky by a short length of elastic. Clearly, Skylarks move those of artistic temperament: poets to versify, but nicest of all, perhaps, the composer Ralph Vaughan Williams has captured the very essence of Skylark atmosphere — the wide open skies which are the larks' domain — in *The Lark Ascending*. For each of us, the cheerful but rather repetitive song may lack something in sheer bird musicianship, but more than compensates by what it offers as *the* sound to relax to in summer.

Skylark *Alauda arvensis*

SNAKE-BIRD

The snake-birds — or alternatively anhingas or darters — are a group of waterbirds closely related to the cormorants and found right around the world in tropical freshwater lakes and swamps. They get the name 'snake' from their habit of swimming with their body submerged and only the slender neck, head and dagger-like beak showing above the water, giving much the same appearance as a swimming snake. Amazingly, snake-birds (like cormorants), although among the longest-standing birds in evolutionary terms and always associated with water, lack effective feather waterproofing! In consequence, to avoid becoming waterlogged and sinking ignominiously, they must clamber out of the water at intervals and stand, wings extended, to dry out. In this posture they seem to be auditioning for an appearance as part of an heraldic coat-of-arms.

African Darter *Anhinga rufa*

SPOONBILL

Spoonbills *Platalea leucorodia*

The spoonbill tribe, though not large, is worldwide in distribution, favouring wetland areas in the tropics or warm-temperature zones with shallow muddy margins where they can hunt for food. Though worldwide, their plumages vary from rich cerise, as in the European Spoonbill, to plain white, but each species possesses the long spatulate beak from which the name is so obviously derived. The 'spoon' end of the beak is richly equipped with nerve endings, making it into a very sensitive food-locating and -identifying organ. Much of the diet is shellfish and small crustaceans lurking in the ooze (and obviously located by touch), but fish also are often taken. Working as a group, the spoonbills herd a school of fish into the shallows, keeping 'in touch' with their prey as their beak tips scythe through the water. Only when their fast-moving intended target doubles back between their legs does this system give spectacularly hilarious results!

STARLING

The Spotless Starling of the Mediterranean, like its Eurasian relative the Starling, must be one of the few birds as familiar to office-workers in cities as it is to countrymen. Many city statues and window-ledges support roosting starlings in considerable warmth (3° warmer than in the country) and the half hour before a winter sunset will see wheeling mobs circling over the city squares. Once perched, they do not 'switch out the light and go to sleep' but continue to chatter, a noise that has given rise to the collective noun — very picturesque — for such gatherings: a *murmuration* of starlings. Cavities under the eaves provide plentiful nest sites. When they get to their food, be it in the fields or in the town parks and gardens, Spotless Starlings continue to be noisily gregarious. There is much apparent aggression, with birds shaping up to each other with feathers bristling and little hops of anger into the air. Although they may seem to be greedy bullies, their technique has to be admired and they are *very* successful birds.

Spotless Starling *Sturnus unicolor*

STEAMER-DUCK

Falklands Flightless Steamer-ducks
Tachyeres brachypterus

The three steamer-duck species are confined to the coasts of the southern tip of South America and the island groups between Cape Horn and the Antarctic, in particular the Falklands and South Georgia. All are heavily built, and in many ways parallel the eiders of the Northern Hemisphere. For example, the massive structure of the wedge-shaped beak is well equipped with muscles to give good crushing power, and, like the eiders, this allows them to cope with the shellfish that they obtain by diving into the kelp beds just offshore. Two of the three steamers, including the Falklands species, are actually flightless, and all three tend to escape danger by swimming strongly away, sometimes using their wings to assist progress. This creates a regular pattern of splashes, giving much the same impression as the old-time Mississippi paddle steamer — hence the name 'steamer' duck.

STINKER

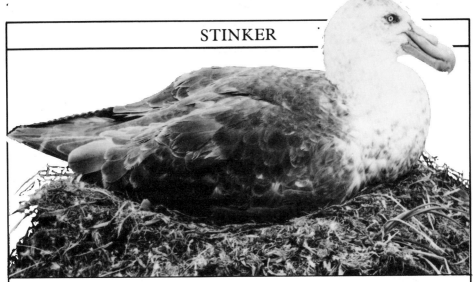

The Stinker, or Giant Petrel, was also known as 'Nelly' (though for what reason is unknown) by the seafarers of the past. It is the largest of the petrel family, and inhabits sub-Antarctic waters, nesting on remote islands and being one of only a few petrels habitually nesting above ground. Out of the breeding season, Stinkers circumnavigate the globe in the prevailing westerly winds, penetrating north almost to the tropics. Supremely elegant and efficient flying out at sea, the Stinker is one of the least graceful and most unattractive birds at close range in both appearance and habits. Out at sea it feeds on squid and shrimps, but near its nest it is a predator on penguins and other birds. At all times Stinkers will flock to sewage outfalls, refuse or carrion, brutally elbowing their way into a throng of birds and gorging to such an extent that they may not be able to take off again — a sight best and most horrifically seen near whaling boats or factories.

Giant Petrel *Macronectes giganteus*

STORK

White Stork *Ciconia ciconia*

Though decreasing over much of their breeding range in western Europe, enough White Storks remain in central Europe and Asiatic regions to create the vast flocks that gather each autumn to cross the Bosporus on the southward migration into Africa for the winter. Huge flocks accumulate near Istanbul and, as the temperature rises, take to the air and ascend in crowded spirals, often thousands strong, to glide over the short sea-crossing. Tree nests are widespread, but the White Stork is best known for its habit of building huge nests of twigs and branches on the roofs and towers of larger houses and churches. To have nesting storks is considered an honour to the house in most countries: hence perhaps the origins of the legends that storks bring the new-born babies to the house.

SUNBIRD

The sunbirds, brilliantly variegated avian jewels darting from flower to flower, are often compared with the hummingbirds of the New World. True they share the same dazzling plumages, which owe much of their glory to metallic iridescent sheens, and the sunbirds do also often feed on nectar. Most, like the Palestine Sunbird, have long, often curved and finely pointed beaks that allow them to reach the nectar in tubular flowers, or to pick up the small insects that are their other major source of sustenance. They are, though, not even distantly related to the hummingbirds; although they are fast fliers, close observation shows that they lack the extreme sophistication of the hummingbirds' flight, which allows hovering and even flying in reverse when feeding on the wing. More often than not, sunbirds perch to feed (though they can and do hover) and, as a reflection of this habit, they are equipped with quite normally sized legs and feet. These are in marked contrast to the hummingbirds' tiny appendages, which gave rise to the name of their order 'Apodiformes', which also includes those other aerial birds the swifts and means footless. Sunbirds occupy the tropical areas of the Old World, with most species in Africa and Asia: the Palestine Sunbird (often called the Orange-tufted) is one of the most northerly, occurring as its name implies in the arid scrublands at the southeastern corner of the Mediterranean.

Female Palestine Sunbird *Nectarinia osea*

SWALLOW

'One Swallow does not a summer make', the saying goes. True enough, but how much pleasanter summers are for the presence of Swallows around our homes. Perhaps the saying more reflects the fact that, although the occasional Swallow may arrive back late in March, the bulk of its fellows sensibly wait until April is established and warmer weather has provided adequate insect supplies to keep them fed. Swallows have an amazing migration, travelling north each spring, from wintering areas in the extreme south of Africa. A prodigious journey, carried out at high speed, but at least Swallows, as aerial feeders, can refuel in mid-air! Today we put numbered light-alloy rings on birds' legs to help trace their migratory routes and establish journey times, but marked Swallows were used as far back as the Punic Wars in 300 BC in the role of carrier pigeons. Adult Swallows were taken from their nestlings and smuggled out of besieged garrisons; they were later released, to 'home' to their nests, with a message attached to their legs giving details of the relief force's progress. Shortly after, to defeat the 'bookies' of the day, the results of chariot races in Rome were conveyed to Volterra — 120 miles (193 km) away — quickly and effectively by tying threads of the winners' colours to the Swallow's leg.

Swallow *Hirundo rustica*

SWAN

Mute Swans *Cygnus olor*

Mute Swans are among the contenders for the title of heaviest flying bird, and they are gracefully familiar birds right across Europe. Despite their name, the word 'swan' is etymologically related to the Anglo-Saxon stem from which 'sound' and 'sonnet' are also derived. Anyway, they are far from mute: the creaking music of their wings in flight is one of the wildest of marshland sounds, emphasising their power. This power, too, is reflected in the hissing and honking grunts of a male — cob — swan, wings raised like a galleon in full sail, powering towards you through the water, pushing a bow wave before his breast. Stories of human limbs being broken by a single blow of the wing pass through the birdwatcher's mental imagery at such times as this. In Britain, Mute Swans are connected with the Queen and the ancient Guilds of the City of London — most of those on the River Thames are 'owned' by them. For ordinary citizens, the connections are strong, too. In a survey by the late W.B. Alexander, over one third of the public houses with names associated with birds were called 'The Swan'.

TANAGER

The very word 'tanager' conjures up tropical America in any birdwatcher's mind. This large family, almost 200 species strong, is typical of that region. They are remarkably uniform in size, ranging from 4 to 8 inches (10–20 cm), and are often said to contribute more than any other family to the brilliant colours displayed by South American birds. Hummingbirds may be numerous and spectacularly colourful, but their full beauty can be appreciated only under rather special lighting conditions, while the rainbow hues of the tanagers are there for anyone to see, given that he has the patience to watch the treetops until they appear. Like a number of other South American groups of birds, toucans, Hoatzin and macaws for example, the tanagers have inherited their name from the Tupi Indians living in the Amazonian forest of Brazil, who are obviously keen birdwatchers with excellent field skills. Tanager colours genuinely span the rainbow, with optical effects like iridescence thrown in as a bonus. Sometimes it seems that most of the colours are occurring on one bird, as in the Paradise Tanager, or in Gould's Tanager (known only from *one* specimen sent in a shipment of skins to England in Victorian times) or Arnault's Tanager (of which again there is only one specimen, which reached Paris alive as a cagebird during the nineteenth century!). This is Mrs Wilson's Tanager.

Mrs Wilson's Tanager *Tangara nigrocincta franciscae*

TERN

The terns, aptly given the popular name 'sea swallows', are perhaps the most graceful and elegant of our seabirds. All are migrants, travelling to fish in the oceans of the Southern Hemisphere during the winter, returning to our coasts to breed in summer. Sandwich Terns, named after the coastal town in southeast England, are colonial breeders, and the colonies may be thousands strong. Eggs and young alike are perfectly camouflaged against the background of sand flecked with fragments of seaweed and shell. Although the colonies seem close-packed, terns are vociferously abusive neighbours and squabbles are common, so closer inspection will reveal that nests are always at least two beak-thrusts apart! During the breeding season terns fish close inshore, flickering along lazily in the air a few feet above the waves before suddenly turning and plunging headlong into the sea. Often these dives result in an audible 'plop' and a considerable splash, but the bird penetrates only a few inches rather than submerging deeply. The prey is usually small fish, and these are taken one at a time back to the colony. Early in the season, the male will often bring a fish back to his mate as she stands beside the shallow scrape which serves for a nest.

Sandwich Terns *Sterna sandvicensis*

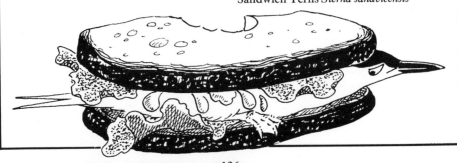

THICKNEE

The thicknees, or stone curlews, are slightly extraordinary members of the huge worldwide family of waders. There is one species in Europe. Rarely are they to be seen beside the sea or feeding on sheltered estuary mudflats, and even during the breeding season their chosen habitat is far from the marsh or tundra favoured by most of their relatives. They are birds of arid landscapes, where stony soil and stunted vegetation intermingle. Largely nocturnal (and with an unearthly yodelling cry), they have huge, fixedly staring yellow-rimmed eyes the better to see their insect prey. The colloquial name 'thicknee' is derived from the knobbly knees (actually, in true anatomical terms, their ankles) so conspicuous on their long legs. At no time are these more noticeable and arthritic-looking than when an incubating bird stealthily returns to her nest. She pauses over the eggs, then with apparent acutely painful slowness folds her legs and lowers her body down onto them. Meanwhile, her watchful head has remained aloft — the neck stretching as the body sinks — until it too is lowered, like the periscope of a submarine.

Stone Curlew *Burhinus oedicnemus*

TINKER-BIRD

Smartly, if rather exotically, elegant, this tiny tinker-bird, which is smaller than a Blue Tit, belongs to the barbet family. Barbets are most abundant in Africa (though they do occur in Asia and America too) and the Golden-rumped Tinker-bird lives in the forests of East Africa. Despite their small size, they are robustly built birds, with a typically solid and oversized barbet beak with a serrated edge. Although ostensibly fruit-eaters, taking the occasional insect, tinker-birds will inflict painful bites to the fingers of an incautious bird-ringer.

These are noisy, quarrelsome birds, moving in flocks, which seems strange in the light of the frequent fights which break out among them. They are relatively poor fliers, with short rounded wings reminiscent of a minute woodpecker. Their feet, too, are woodpecker-like, with two powerful toes pointing forwards and two back, giving the most effective grip on the bark. Their name comes from their call, a 'tink . . . tink . . .', inescapable if you are in tinker-bird country.

Golden-rumped Tinker-bird
Pogoniulus bilineatus

TIT

Though the black cap of the Coal Tit may not be so shiny as a lump of coal, the likelihood is that the name was coined (along with the Scottish version, Coaly-Hood) back in the days when the common fuel would have been the much greyer, less glossy, charcoal. Coal Tits are conifer specialists, climbing agilely among the cones and twigs, seeking seeds or insects as food. Despite this, they have taken readily to visiting garden bird tables, where they join with the common Great and Blue Tits in enjoying nuts and fat. Peanuts have a particular attraction for them, and bring out their urge to cache food against future times of shortage. Thus the Coal Tit will fly with peanut after peanut and hide them, often in a nearby yew tree, only for a cunning Blue Tit, watching the operation, to dart in and remove the nut as soon as its concealer has departed!

Coal Tit *Parus ater*

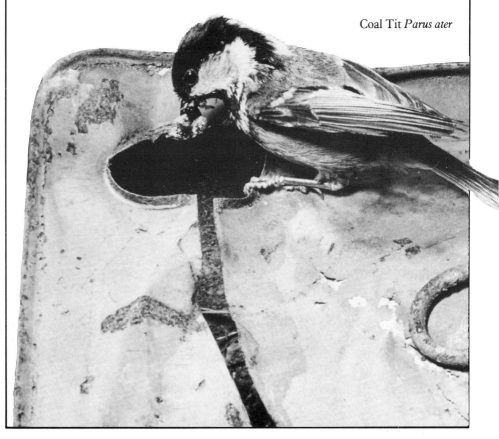

TOUCAN

Toco Toucan *Ramphastos toco*

Toucans range from Mexico south to Paraguay, with their centre of abundance probably somewhere in the unexplored tracts of the Amazon rain forest. The Toco Toucan lives in north and east South America. Knowledge of the toucan, however, has penetrated throughout the civilised world not just because they are often kept in captivity but more because, in those areas where alcoholic beverages are popular, *Guinness*, the brewers of a unique type of stout (a dark ale, or porter), have adopted the toucan as an emblem. *Guinness* stout has a worldwide market, relying heavily over the years on advertisements featuring the toucan to bring it to the attention of drinkers. As with *Guinness*, young toucans take several weeks to mature. In their hole nests, the young chicks have short, wide flattened beaks, with the lower mandible longer than the upper, and a thick fleshy tongue completely filling their mouths. As they mature, so their beak grows more like the adults'. As another adaptation to the long fledging period that fruit-eaters must endure, toucans develop 'knee' or 'heel' pads (actually situated on the ankle joint in the middle of the leg) to save undue wear and tear as they squat to take the weight off their feet on the unlined floor of the nest: heavy drinkers might envy them!

TREECREEPER

Treecreepers are small brown fluffy birds, spending much of their time, as the name suggests, creeping up trees. Up it always is, as Treecreepers (like woodpeckers) have specially strong tail feathers that serve as a shooting-stick-like prop. This aids the bird as it seeks food — usually insects or their eggs concealed in crevices in the bark — peering with large eyes protected by ferocious-looking beetling eyebrows. During display, they spiral up trunk after trunk — rather like maypole dancing. During the winter months, many Treecreepers seek out wellingtonias, soft-barked trees related to the giant redwoods. Here they shelter overnight in hollows the size and shape of an egg scooped out of the bark. Although wellingtonias were introduced to Europe only a century ago, and then as rather unusual trees, this habit quickly became widespread.

Treecreeper *Certhia familiaris*

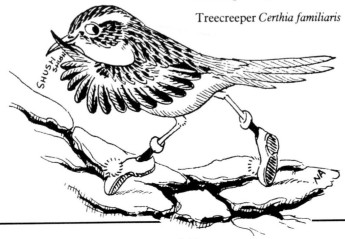

TRUMPETER

White-winged Trumpeter
Psophia leucoptera

Deep in the wet lowland jungles of the northern areas of South America lives an extraordinary family of birds — three species in all — which show many characteristics resembling those of the rails and crakes, and many others resembling (astonishingly) those of the cranes! But this is to the taxonomist. To the uninitiated, the White-winged Trumpeter looks like a bulky black-and-white chicken on stilts. The name 'trumpeter' is derived simply from the birds' strident calls, but it is in their association with man that trumpeters are at their most enigmatic. Trumpeters are sociable flocking birds, moving about on the forest floor in large flocks and so unwary as to be described as positively dim. They fly poorly, preferring to run, but oddly can swim quite well. They make very good eating, and because of their simplicity are easily caught — hence few are now to be found near human settlement. Quixotically, they are easily tamed, and many Indian households keep them with their chickens, as with their loud voices they make excellent watchdogs!

TURNSTONE

As its name suggests, the Turnstone seeks much of its food beneath stones. A wader, most familiar on rocky coastlines, it also uses its beak (unusually flattened from top to bottom) like a miniature shovel as it seeks small worms, shellfish and shrimps beneath the lines of rotting seaweed on the strandline. Turnstones have been recorded feeding in a variety of unusual and opportunist circumstances. A small flock was recorded feeding in the rainwater guttering of a large fish warehouse, doubtless finding scraps dropped by scavenging gulls. Another was observed feeding inside the carcase of a dead sheep washed up on the beach in England and one was seen pecking away at a dead wolf in Canada. But most macabre of all is the record of several feeding on the long-dead corpse of a drowned human, washed ashore in Wales.

Turnstone *Arenaria interpres*

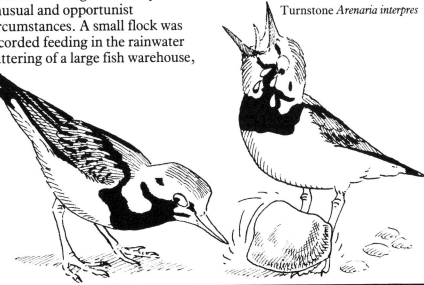

VULTURE

Among the smaller and more slimly built of the vulture family, the Egyptian seems to hold its own at the macabre feasting that goes on whenever a suitably rotting carcase is located, by its nimbleness as it

Egyptian Vulture *Neophron percnopterus*

darts between its larger relatives — even wriggling under them on occasion — snatching a sizeable morsel and vanishing before the larger birds can retaliate. Young Egyptian Vultures are dark brown, but the adults are more or less white, and so tend to show more obvious signs of their gruesome profession in terms of bloodstained plumage. They have one remarkable attribute. When birds conquered the air back in evolutionary time, they 'sacrificed' (in anthropomorphic terms) their 'hands'. As one consequence of this, extremely few birds use tools of any sort, but the Egyptian Vulture is an exception. It is a lover of eggs, but at a loss when confronted by an Ostrich egg with its very robust shell. To overcome the problem, Egyptian Vultures will pick up nearby chunks of rock in their beaks and use these to smash the shell, to reach the amazing 2-litre (3½-pint) volume of potential food within.

WARBLER

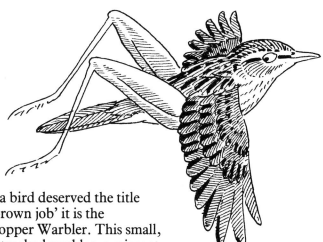

Grasshopper Warbler *Locustella naevia*

If ever a bird deserved the title 'little brown job' it is the Grasshopper Warbler. This small, black-streaked warbler, a migrant to Europe from winter quarters in Africa, is the ultimate in skulking and generally retiring birds. It lives and breeds in dense, often damp or marshland vegetation, but is now showing some adaptability in colonising freshly planted conifer plantations. The song is best described as a continuous churring reel — rather like the noise of fishing line being pulled from the reel on a rod — often continuing for minutes on end and produced both day and night. Gilbert White, writing in the eighteenth century, commented that a Grasshopper Warbler 'will sing at a yard distance, provided it is concealed'. This is true, and the songster is uncommonly difficult to locate even at this range: the bird sings, throat bulging, mouth wide, turning its head from side to side and giving a ventriloquial effect. Though audible at up to a kilometre — just over half a mile — the song is a severe test for the onset of old age in birdwatchers: when the Grasshopper Warbler's high frequency eludes the hardening ear-drums, then beware!

WAXWING

Waxwings — beautiful birds by any standards — breed in the circumpolar belt of open birch and conifer forest that merges into the Arctic tundra. For much of the time

Waxwing *Bombycilla garrulus*

they are remote from man, but when winter food shortages force them south they have shown a remarkable predilection for the berried shrubs commonly grown in gardens. Here they will feed, showing little fear of man, using parrot-like antics and agility to reach *Cotoneaster* berries or crab apples. The name 'Waxwing' is obviously derived from the red sealing-wax-like blobs on the end of some of the wing feathers, but a couple of hundred years ago the Waxwing was known as the 'Bohemian Chatterer': Bohemian because of its exaggeratedly colourful plumage, 'chatterer' because Waxwing flocks are for ever keeping in touch with their trilling calls. Chatterer is reflected in their specific name *garrulus*, and in North America the bird is still known as the Bohemian Waxwing.

WEAVER

The rather pompous-sounding name of Napoleon Weaver is an alternative to Golden Bishop, and is applied to one of the more colourful of the large African weaver-bird family. Napoleon Weavers may well get their name not from their elegant black-and-gold uniform but from their abilities as 'little corporals' martialling the flocks under their control. Each male has a harem of females, each on a nest in his grassy East African plains territory. Around this bevy of wives the male will circuit, buzzing fussily and angrily like a slightly damp firework or a yellow balloon that, once inflated, has slipped from your fingers. Other weaver-birds are conspicuous because of their neatly woven predator-proof hanging nests, yet others because they are damaging to crops. One, the Quelea, may well be the most numerous bird in the world. It is a severe pest of cereal crops, and is often the subject of draconian control measures. In the Republic of South Africa, in one year alone, over 100 million Quelea were destroyed without materially reducing the level of crop damage! Nor was the Quelea population much affected.

Male Napoleon Weaver *Euplectes afer*

WHALEBIRD

Antarctic Prion *Pachyptila desolata*

Known to old seafarers, especially whalers and sealers, as whalebird or icebird, the prion (to give it its proper name) was also called firebird. This strange title arose from its tendency (and that of many other nocturnal seabirds) to be attracted to bright lights or fires near the beach. Prions are typical small petrels, with a short 'tube-nose' on the ridge of the beak, nesting in burrows on many of the islands surrounding Antarctica. Unlike many other petrels, however, they regularly gather and fly in flocks, travelling fast enough even to dodge skuas. They fly low over the waves, frequently and suddenly banking and turning in precision unison. Like whales, prions are plankton feeders, loving krill, so whale and whalebird are often seen in company. Like whalebone whales, too, they scoop up a throatful of water and krill 'soup', and then expel it through filter-like combs (which collect the krill) on the side of the beak, using their thick fleshy tongue like the plunger of a pump.

WHEATEAR

The white rump and sides to the upper part of the otherwise almost black tail are a most characteristic feature of the Wheatear as it flits away low over the ground. It is from this feature that the Wheatear gets its name, which has nothing at all to do with ears of wheat. 'Wheat' is a corruption of the Anglo-Saxon word 'hwit' for white, and 'ear' an abbreviation of again the Anglo-Saxon word 'earse', which refers to the anatomical area in humans more politely called the bottom, which approximates to the rump of a bird. Wheatears are birds of stony open ground, or close-cropped turf swards, often nesting below ground. The Greenland race, nesting up into the Arctic Circle, has one of the most prodigious migrations of small birds. To reach their wintering grounds in Africa, they may have a 12,000-kilometre (7,500-mile) journey, and some include in this a non-stop trans-oceanic hop of about 3,000 kilometres (1,800 miles) direct from Greenland to Spain!

Male Wheatear *Oenanthe oenanthe*

WHIPBIRD

It is easy to discover how the Eastern Whipbird got its name: it comes from the thick undergrowth of the rain forest and eucalyptus forests of eastern Australia, and its call sounds exactly like the crack of a whip. But that is the end of easy dealings with the whipbird. It is a recluse, keeping its elegant, black-and-white, crested head well down in the bush, and though frequently heard is tantalisingly much less often seen. It comes, too, from a strange and poorly understood family, the rail-babblers. With their relatively robust beaks, strong legs, long tails and generally rather scatterbrained behaviour, they resemble some of the true babblers, but other features do not support this. Other members of their family include the Chirrupping Wedgebill, whose call is written 'tootsie, cheers!', and the Chiming Wedgebill, whose call is usually transliterated as 'did you get drunk?' It is appropriate to note that both come from dry areas of Australia!

Eastern Whipbird *Psophodes olivaceus*

WHISTLER

The whistlers are a compact subfamily of the Old World flycatchers, called the 'Pachycephalinae', or 'thick-heads'. This scientific name is based more on their relatively large, rounded heads (set on thickset bodies) than on any elephantine resemblances or, indeed, on any lack of mental agility on behalf of the whistlers. Most are birds of scrub or secondary jungle with a well-developed undergrowth, where their main food is a wide variety of insects caught more often by gleaning off the bark and foliage than in typical flycatcher aerial sortie. These are birds of southeast Asia, Australasia and the neighbouring Pacific Ocean islands, and when not breeding they forage in groups with various other undergrowth birds. In the breeding season pairs are territorial and remain in constant contact, maintaining this by regular duetting whistles (or antiphonal calls), male and female calling in immediate succession. Though their habitat is not good birdwatching country, whistlers are curious birds and readily emerge from cover to check on the identity of a birdwatcher mimicking their calls, allowing good views.

Rufous Whistler
Pachycephala rufiventris

WHISTLING-DUCK

The scientific name for the genus of whistling- or tree-ducks is *Dendrocygnus,* which translated means 'tree-swan' — a clear indication of the difficulties faced by early taxonomists in deciding just what sort of birds these unusual waterfowl are. The name refers to their long neck (and upright stance) and to their habit of perching in trees. The whole genus consists of neat and fast-flying ducks of tropical fresh waters and swamps, and all have the delightful clear fluting whistle of a call that gives them their alternative name. Most of the whistling-ducks have fairly restricted distributions, but the White-faced is widespread and will be met with by birdwatchers in the tropics of South America and Africa. This is an unusually broad distribution, but falls a clear second to that of the Fulvous Whistling-duck, which occurs in surprisingly widely separated areas of both North and South America, Africa, India and the Malay archipelago.

White-faced Tree-ducks *Dendrocygna viduata*

WHITE-EYE

The Indian White-eye is one of the astonishingly widespread group of white-eyes, tiny, fine-beaked birds with no obvious relatives. Their generic name is *Zosterops* and, for all their uniformly green above, yellow or white below, plumage and striking white eye-ring, each region and sometimes almost each island in the tropics seems to have developed 'its own' peculiar *Zosterops* species. White-eyes are birds of trees and bushes, rarely descending to the ground. They feed on insects, and especially on nectar and fruit. Nectar is obtained from flowers with the aid of a highly adapted brush-like tongue. Fruit, which is attacked to the extent that tropical fruitgrowers sometimes regard white-eyes as a pest, is approached in a similar way: the sharp beak is used to puncture the skin, then the brush tongue is used to extract the pulp and juice from within.

Indian White-eye *Zosterops palpebrosa*

WHITETHROAT

Rather more of a woodland bird than its cousin the common Whitethroat, the Lesser Whitethroat is a skulking bird with a brief monotonous song, produced from deep in cover and quite unlike the Whitethroat's scratchy but jaunty song flight high above the hedgerows. Perhaps this skulking behaviour, and the dark, highwayman's-mask-like patches over its eyes, have contributed to the colloquial vernacular name for the Lesser Whitethroat — 'mountebank' — quack or deceiver. Catastrophe overtook many of western Europe's Whitethroats in the late 1960s, when an extended drought in the Sahel zone of West Africa reduced their wintering grounds to a dust bowl, starving humans, cattle and birds alike. The Lesser Whitethroat escaped by virtue of its unusual migration route: southeast through Italy in autumn to winter in the Sudan, north and then northwest through Syria on the return journey in spring.

Lesser Whitethroat *Sylvia curruca*

WHYDAH

Male Pin-tailed Whydah *Vidua macroura*

The whydahs are part of the galaxy of finch-like birds that cover Africa south of the Sahara Desert, ranging from the true weavers to the weaver-finches — the whydahs falling somewhere between the two. The whydahs are fascinating in that they are parasitic — perhaps 'social parasites' would be the best descriptive term. They occur in flocks, with males outnumbered considerably by females and immatures. Each whydah species seems to have evolved as a nest parasite of one of the other finches. In the case of the Pin-tailed Whydah, the females lay in the domed grassy nests of the Grey Waxbill. The eggs themselves, usually two or three in each 'foster' nest, closely mimic in colour those of the host, but, most remarkably, the markings inside the mouth of the nestling are an exact replica of those of the host chick. This is vital if the parasites are to be fed, as the markings are the visual clue stimulating the homecoming parent to feed a nestling.

WIDOW-BIRD

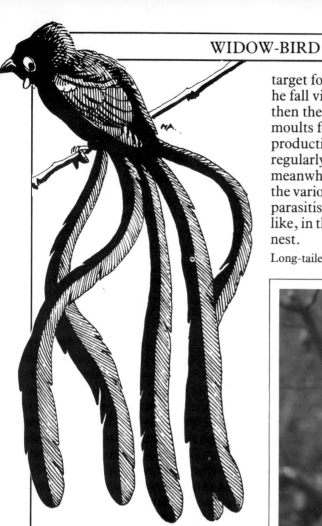

Outside the breeding season, widow-birds roam the open African plains in groups of streaky-brown, uniformly nondescript bunting-sized birds. Following the rains, when richer food supplies are available, a transformation takes place. The flock is composed of a number of females and several males, of which only the most senior (or dominant) moults. A greater contrast with his camouflage browns could hardly be imagined as he dons his ornate long-tailed full breeding plumage. His flight is much slower because of his elaborate tail, and so complex is his mid-air display that he is an easy target for any bird of prey. Should he fall victim to a hawk or falcon, then the next male in the hierarchy moults for his short but hectic productive life. The females, so regularly 'widowed', are, meanwhile, busy locating pairs of the various waxbill finches that they parasitise, laying their egg, cuckoo-like, in the unwitting foster parent's nest.

Long-tailed Widow-bird *Euplectes progne*

WILLIE WAGTAIL

Willie Wagtail *Rhipidura leucophrys*

'Willie Wagtail' was the popular name first given by homesick settlers in Australia to a black-and-white bird in some ways resembling the Pied Wagtail of their British homeland. The Willie Wagtail, though, wags its tail from side to side, almost without ceasing, not up and down like the true wagtails. It is the largest of the Australasian flycatchers, with the broad, flat, hooked beak, surrounded by bristles, typical of that family. These long-tailed flycatchers form a separate group, the fantails, which earned their name from their constant habit of flicking the tail open and shut, like a fan. It is thought that this action, with the tail's striking pattern, helps to frighten resting insects into flight, when the fantail darts out from its perch to catch them. The supremely neat nest is of fine grasses, bound together and anchored with numerous spiders' webs to the merest protrusion from the bark on a branch.

WOODPECKER

Although one of the smaller woodpeckers, little more than sparrow-sized, the Downy is both one of the most numerous and one of the most widespread of North American woodpeckers. The feet are powerful, the toes arranged two forward, two back, to give the best grip on the bark, and the inevitable head-uppermost posture results from the use of the tail, with its specially strengthened central feathers, as a shooting-stick-like prop to support the feeding bird. The beak is relatively small but functions as a combined hammer and chisel, both in excavating a nest hole within the tree and in extracting insect food. A pad of cushioning tissue between base of beak and skull may help to prevent continuous headaches! The tongue is long, and has a barbed horny tip which can be used to 'harpoon' and then withdraw grubs which would otherwise be out of reach.

Downy Woodpecker *Dendrocopos pubescens*

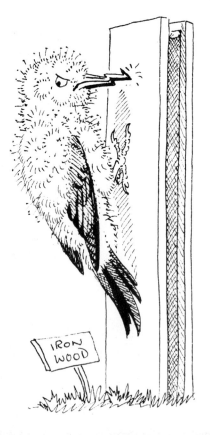

IRON WOOD

WRYNECK

Although closely related to the other members of the woodpecker family and possessing the same powerful dagger-shaped beak and long tongue for extracting insects from their burrows, the Wryneck is very much a 'one-off' bird. It spends much of its time on the ground, as ants are a favourite food, and has a soft-feathered tail like most normal birds, not a rigid prop like the true woodpeckers. Nor is it brightly coloured like most woodpeckers: the Wryneck is superbly camouflaged in a mixture of vermiculated browns, buffs, greys and blacks. But strangest of all are its antics, some related to display, others difficult to explain. Writhing neck movements and hissing calls as the Wryneck creeps around give rise to its name, and remind the watcher of a snake — hence the colloquial 'snake bird' — moving in grotesque slow motion.

Wryneck *Jynx torquilla*

YAFFLE

The gloriously coloured Green Woodpecker, more often appearing golden as it swoops across the meadows, has the colloquial name 'Yaffle' in many areas because of its ringing cry, which sounds like peals of slightly hysterical laughter. More terrestrial than most woodpeckers, Green Woodpeckers specialise in eating meadow ants. They can cut into ant hills with their beaks and then extend their tongue, which is several inches long and covered in sticky saliva, along the ant tunnels, dragging out the hapless insects and eating them. Sometimes they do exert their wood-pecking capabilities, but to man's detriment by attacking cedar-wood shingles on houses and even on church steeples! These attacks occur in winter, and the woodpeckers, rather than attempting to disrupt the sermon or compete with the bells, are seeking blowflies hibernating in the sheltered cracks behind the shingles.

Green Woodpecker *Picus viridis*